Home Office/The Scottish Office

GUIDE TO FIRE PRECAUTIONS IN EXISTING PLACES OF ENTERTAINMENT AND LIKE PREMISES

LONDON: The Stationery Office

Sixth impression 2000

ISBN 0 11 340907 9

NOTICE

This guide does not take account of amendments to fire safety provisions which have been implemented since its publication. These include changes to British Standards and the introduction of the Fire Precautions (Workplace) Regulations 1997 (as amended). These regulations cover fire precautions in most workplaces and require a risk assessment to be carried out either by the employer or at their request. In order to obtain a full understanding of current requirements this guide must be read in conjunction with the relevant sections of "Fire Safety An employer's guide" [ISBN 0 11 341229 0] published in 1999.

Contents

		Page
	Members of the Working Party	3
	Acknowledgements	4
	Introduction	5
	Definitions of Terms used in the Guide	8
CHAPTER 1.	MANAGEMENT RESPONSIBILITIES	13
CHAPTER 2.	ACCESS AND MEANS OF ESCAPE FOR DISABLED PEOPLE	20
CHAPTER 3.	CONSTRUCTION, CLASSIFICATION AND SITUATION OF BUILDINGS	24
CHAPTER 4.	FIRE RESISTANCE AND SURFACES OF WALLS AND CEILINGS; FLOOR COVERINGS, FURNITURE AND FURNISHINGS	29
	4.1–4.2 Fire resistance	
	4.3–4.5 Surface finishes on escape routes	
	4.6–4.7 Surfaces of walls and ceilings etc	
	4.8–4.10 Fire/smoke spread	
	4.11 Advice on floor coverings	
	4.12–4.19 Advice on furniture and furnishings, and synthetic materials	
	4.20–4.23 Artificial and dried foliage	
	4.24–4.27 Cellular foam	
CHAPTER 5.	MEANS OF ESCAPE	38
	5.5–5.7 General principles	
	5.9 Accommodation of children in places of entertainment	
	5.10–5.14 Distance of travel	
	5.16–5.20 Occupant capacity	
	5.21–5.26 Exit capacity	
	5.27–5.44 Stage 1 — Travel within rooms	
	5.45–5.53 Stage 2 — Travel from rooms to a stairway or final exit	
	5.54–5.78 Stage 3 — Travel within stairways to a final exit	
	5.94–5.110 Doors and fastenings on doors	
	5.111–5.123 Exit and directional signs, notices	
CHAPTER 6.	ELECTRICITY, GAS AND OIL	69
	6.1–6.4 Electrical, gas and oil equipment not to give rise to danger	
	6.6–6.26 Electrical safety	
	6.27–6.51 Lighting	
	6.52 Inspection of electrical installations	
	6.54–6.58 Heating appliances	
	6.59 Exclusion and admission of the public	

Page

CHAPTER 7. SEATING ARRANGEMENTS 80

CHAPTER 8. REQUIREMENTS IN PREMISES HAVING A STAGE 90

CHAPTER 9. MEANS FOR GIVING WARNING IN CASE OF FIRE 103

CHAPTER 10. FIRE FIGHTING EQUIPMENT 109

CHAPTER 11. USE OF PYROTECHNICS, SPECIAL EFFECTS AND FIREARMS 117
- 11.1–11.25 Pyrotechnics
- 11.26–11.38 Smoke and vapour effects
- 11.39 Real flame
- 11.40–11.44 Lasers and strobes
- 11.45–11.46 Firearms

CHAPTER 12. FIRE PRECAUTIONS IN SMALL PREMISES 134
- 12.5 The manager's responsibilities
- 12.6 Premises managed by voluntary organisations
- 12.7–12.12 Attendants
- 12.14–12.20 Construction and fire separation
- 12.22–12.38 Exits, stairways, doors etc
- 12.39–12.46 Exit and directional signs
- 12.47–12.51 Notices on doors

CHAPTER 13. SPECIAL PROVISIONS FOR TEMPORARY STRUCTURES AND PLACES OF ENTERTAINMENT WHICH ARE UNDER COVER IN OTHERWISE OPEN AIR SITUATIONS 147
- Definitions
- 13.2–13.5 Site selection and access
- 13.7–13.8 The manager's responsibilities
- 13.9–13.13 Attendants
- 13.18–13.20 Tented structures
- 13.21–13.27 Air supported and pneumatic structures
- 13.30–13.57 Means of escape for enclosed or covered structures
- 13.68–13.77 Seating and gangways
- 13.82–13.87 Other fire precautions
- 13.88 Additional provisions for circuses and like events involving animals

APPENDICES

A. Surface Finishes of Walls, Ceilings, Partitions etc 164

B. Use of Batteries 166

C. British Standards References 168

D. Reference Sources 173

INDEX 176

MEMBERS OF THE WORKING PARTY

Chairman:
Mr T Greenwood (HM Fire Service Inspectorate, Home Office) (1983-1987)

Vice/Co-Chairman:
Mr D Wilson (HM Fire Service Inspectorate, Scottish Home & Health Department) (1983-1989)

Co-Chairman:
Miss V Jenson (Home Office) (1987-1989)

Members:
Mr A Gatfield (HM Fire Service Inspectorate, Home Office) (1983-1987)
Mr A R Hart (Fire Service College) (1983)
Mr D G Wright (London Fire Brigade) (1983)
Mr A Langdown (West Midlands Fire Service) (1983-1989)
Mr P Robinson (Fire Service College) (1984-1989)
Mr J Castle (London Fire and Civil Defence Authority) (1984-1989)
Mr D Coward (London Fire and Civil Defence Authority) (1988-1989)
Mr G Tinley (HM Fire Service Inspectorate, Home Office) (1987-1989)

Secretariat:
Mr R Rice (Home Office) (1983-1987)
Mr D Garland (Home Office) (1983-1985)
Miss C Browne (Home Office) (1985-1987)
Mr P Boshell (Home Office) (1987-1989)
Mr C Dickety (Home Office) (1983-1985)
Mr A Casey (Home Office) (1985-1989)
Mr P Moore (Home Office) (1988)

ACKNOWLEDGEMENTS

The Home Office and Scottish Home and Health Department gratefully acknowledge the assistance provided by all who contributed to the consultation exercise which was carried out during the preparation of this guide. Also acknowledged is the kind permission of the following organisations to reproduce certain photographs, diagrams and drawings in the guide:–

The Association of British Theatre Technicians

Brighton Borough Council

The London Fire and Civil Defence Authority

The London Residuary Body (on behalf of the former Greater London Council)

The Marquee Hirers' Section of the Made-Up Textiles Association

The Village Halls Forum

INTRODUCTION

About this guide

1. This guide recommends standards for means of escape and other fire precautions in a wide range of premises used for public entertainment and recreation. It was drafted by a working party set up by the Home Office and the Scottish Home and Health Department following the fire at the St Valentine's Dance in 1981 at the Stardust Disco in Dublin in which 48 people died. It was completed following extensive consultation with organisations representing owners, occupiers and users, enforcing authorities and professionals.

Who the guide is for

2. *The guide is directed mainly at the professional or technically trained officers in local authorities and fire authorities who are charged with giving advice on fire safety matters, whether to licensing authorities or to managements. It aims to set acceptable standards of safety and encourage consistency of enforcement, while at the same time leaving scope for flexibility and the exercise of professional judgement and common sense.* In interpreting this guide enforcing authorities will particularly consider:–

(a) staffing levels and the quality of staff supervision before, during and after the entertainment;

(b) the limitation of adapted premises;

(c) the frequency of use;

(d) the costs of fire precautions measures and the possibility that alternative solutions may be available which are less expensive; and

(e) the special needs of the disabled (see Chapter 2).

3. *The guide also provides advice for licensees, owners and managers on their responsibilities and the safety levels they may be expected to meet.* It in no way replaces, however, the important need for potential licensees or managers to consult the licensing authority well in advance of any use or entertainment for which a licence may be needed. They should also be present when their premises are inspected so that there is no misunderstanding about the use or proposed use, or about the nature of any fire precautions which are regarded as necessary.

The premises to which the guide applies

4. The guide is provided for existing buildings, that is to say, it is not a design guide for new buildings for which fire safety and means of escape requirements may be contained in Building Regulations, similar national or local legislation, or nationally accepted Codes of Practice such as those published by the British Standards Institution. But once a new building has been brought into use as a place of entertainment, (although the building should be maintained to the standard of fire safety and means of escape provided at the time of construction), much that is in this guide will be relevant.

5. *References to British Standards in this guide are to those Standards in force at the time of publication. Where premises (or parts of premises) were brought into use as places of entertainment at a time when earlier Standards prevailed, those standards should normally be regarded as acceptable until such time as the building is "materially" altered when the later Standard will apply. Where premises (or parts of premises) are brought into use as places of entertainment after the publication of this guide, the references should be construed as references to the latest suitable versions of the relevant Standards when the work is undertaken.*

6. The range of premises for which this guidance is appropriate includes theatres, concert halls, dance halls and discotheques, conference centres and indoor leisure centres. It also provides advice at Chapter 13 on temporary structures, and those places of entertainment which are under cover in otherwise open air situations eg zoos, circuses and open air theatres. It does not apply to sports grounds for which there is a "Guide to Safety at Sports Grounds" revised in 1986 and published by HMSO (ISBN 0–11–341001–8). Neither does it apply to cinemas or other premises used for film exhibitions, to which special regulations apply.

The status of the guide

7. The guide has no statutory force. It is issued without prejudice to the operation of statutes such as the Fire Precautions Act 1971, Building Regulations, the Health and Safety at Work etc Act 1974 or the Safety of Sports Grounds Act 1975, but whenever posssible, the provisions of the guide have been compiled so that they are likely to be consistent with requirements under those statutes.

8. The guide applies to England and Wales and to Scotland but not to other parts of the United Kingdom.

Note: The main statutes under which licences may be required are:–

(a) Schedule 1 to the Local Government (Miscellaneous Provisions) Act 1982;

(b) Schedule 12 to the London Government Act 1963; and

(c) Section 41 of the Civic Government (Scotland) Act 1982;

(which have all been amended by Part IV of the Fire Safety and Safety of Places of Sport Act 1987); and

(d) the Theatres Act 1968;

(e) the Licensing Act 1964 as amended by the Licensing Act 1988; and

(f) the Greater London Council (General Powers) Act 1966, as variously amended.

DEFINITIONS OF TERMS USED IN THE GUIDE

(Defined terms appear in **bold** in the text)

For the purposes of this guide the following definitions apply:–

Acceptable/Accepted means **acceptable** to the licensing authority in writing.

Access room means a room that forms the only escape route from an **inner room.**

Accommodation stairway means a stairway which is provided for the convenience of the occupants in addition to that or those required for escape purposes.

Approval/Approved means **approved** by the licensing authority in writing.

Area of high fire risk means a room or other area which, because of its function and/or contents, presents a greater risk of fire occurring and developing than in the main public areas; such areas may include the main kitchen, the boiler room, store, workshop and dressing rooms.

British Standards include **British Standards** Codes of Practice and Specifications issued by the British Standards Institution.

*Note: Appendix C lists relevant **British Standards**.*

Building includes a temporary or movable building and also includes a permanent structure and any temporary structure, and any vessels remaining moored or on dry land.

Circulation space means a space mainly used as a means of access between a room, or protected shaft, and an exit from the **building** or **compartment**.

Class 0 means a material of limited combustibility which restricts both the spread of flame across a surface and also the rate at which heat is released from it, with a more strict control than Class 1. **Class 0** is not a classification identified in a **British Standard** test and is considered a higher class than Class 1.

Compartment is a part of a **building** separated from all other parts of the same **building** by compartment walls and/or compartment floors.

Dead end means a place from which escape is possible in one direction only, or in directions less than 45° apart which are not separated by **fire-resisting construction**.

Distance of travel means the actual distance that a person must travel between any point in a **building** and the nearest **final exit**; or door to a stairway which is a **protected route**; or door to a **protected lobby** leading to a stairway; or a door for **means of escape** in a compartment wall.

Durably flame-retarded fabric/scenery means fabric/**scenery** that has been chemically treated to render it flame-retardant so that when subjected to the appropriate wetting or cleansing procedure of **British Standard** 5651 it retains a stated performance when tested to **British Standard** 5438 Tests 2A and 2B with a 10 second flame application time in each case.

Emergency lighting means lighting provided for use when the supply to the **normal lighting** fails.

Escape lighting is that part of the **emergency lighting** which is provided to ensure that the escape route is illuminated at all material times.

Final exit means the termination of an escape route from a **building** giving direct access to a **place of safety** such as a street, passageway, walkway or open space and sited to ensure that persons can disperse safely from the vicinity of the **building** and the effects of fire.

Fire door means a door which, if tested under (i) the conditions of test for door assemblies prescribed in **British Standard** 476: Part 22; or (ii) the conditions of test contained in the **British Standard** currently in force at the time of the bringing into use of the **premises** as a **place of entertainment**; or (iii) the conditions of test in the **British Standard** currently in force at the time the door was manufactured; would satisfy the criteria for integrity for 20 minutes or for such longer period as may be specified for particular circumstances. Normally such doors should be rendered positively self-closing.

Fire-resisting construction means the ability of a component of a **building** to satisfy some or all of the criteria specified in **British Standard** 476: Parts 21-24 or the conditions of test contained in **British Standard** 476 in force at the time of the construction or the bringing into use of the **premises** as a **place of entertainment** relating to load bearing capacity, integrity and, where appropriate, insulation, for not less than 30 minutes or for such longer period as may be required in the case of that construction.

Fireworks mean devices or preparations of an explosive or flammable nature used for spectacular effects or display purposes.

Flame-retarded fabric means a fabric which meets the performance requirements of **British Standard** 5867: Part 2: Type B and retains this performance after being subjected to the appropriate wetting or cleansing procedure in **British Standard** 5651.

Heating appliance includes an electric or gas fire.

IEE Regulations for Electrical Installations means the current edition of the Regulations for Electrical Installations published by the Institution of Electrical Engineers (IEE).

Inherently flame-retarded fabric means a fabric which, although not treated to render it flame-retardant, is throughout its thickness a **flame-retarded fabric.**

Inner room means a room from which escape is possible only by passing through an **access room.**

Licensee means the person or persons to whom a current licence has been granted.

Manager means a person of at least 18 years of age who has personal overall responsibility for the management of the **premises** and who may also be:–

(a) the **licensee;**

(b) the owner of the **premises;**

(c) the holder of the franchise for the **premises**;

(d) the hirer of the **premises;** or

(e) a person nominated in writing by any of the above to have responsibility for the **premises.**

Means of escape is the structural means whereby a safe route is provided for persons to travel from any point in a **building** to a **place of safety**, without outside assistance.

Non-combustible material means any material which is deemed to be non-combustible according to the test criteria specified in **British Standard** 476: Part 4.

Non-durably flame-retarded fabric means a fabric that has been chemically treated to render it flame-retardant but when subjected to any of the wetting or cleansing procedures of **British Standard** 5651 that treatment is adversely affected.

Normal lighting means all permanently installed electric lighting operating from the normal supply, ie that source of electrical energy used to provide **normal lighting,** which in the absence of adequate daylight is intended for use during the whole time that the **premises** are occupied. (*Note: Also known as general lighting.*)

Open stage means a **stage** which is not separated from the remainder of the **premises**.

Permitted means **permitted** by the licensing authority in writing.

Place of safety means a place in which a person is no longer in danger from fire.

Place(s) of entertainment means such places to which people resort, including places of recreation, whether as members of **the public**, members of a club or otherwise.

Platform means that part of the assembly space which is arranged for the accommodation of performers and which is not a **stage.**

Premises mean a place, an enclosure, a **building** or that part of a **building** within a place including services or installations used in connection with assembly.

Property (properties, props) means a movable object (other than furniture or **scenery**) used on **stage** during a performance of a play etc.

Protected lobby means a lobby having an adequate degree of protection from fire and forming part or the whole of the horizontal component of an escape route or affording additional protection to an escape route.

Protected route means a route having an adequate degree of protection from fire including walls (other than any part that is an external wall of a **building**), partitions and floors separating the route from the remainder of the **building.**

Protected stairway means a stairway including any exit passageway leading therefrom to its **final exit,** enclosed with (other than any part that is an external wall of a **building**) **fire-resisting construction**.

Public, The: see **The public**

Pyrotechnics are **approved** theatrical devices which contain combustible substances and oxidants which when ignited react either violently or slowly according to their mixture; they include **fireworks**, maroons, flash powder, smoke powder, percussion caps, safety cartridges, firearm blank cartridges, safety fuse and detonators.

Scenery includes any false proscenium, portal, cyclorama or set piece and any curtains or other hangings except the main tabs (house curtains) but does not include **properties**.

Scenery store includes a scene dock, workshop or store other than a small **property** store.

Self-luminous fire safety sign is an **approved** sign which has letters or graphics illuminated by radio-active means and of which the useful life (according to the date indicated on the sign by the manufacturer) has yet to expire and accords with the provisions of **British Standard** 5499: Part 2.

Separated stage means a **stage** which is separated from the remainder of the **premises**.

Separating wall means a wall separating **buildings**.

Stage means that part of the **premises** which is constructed or arranged for the performance of stage plays or entertainments of the like kind.

Stage area means (in addition to the **stage** itself) any scene dock, workshop, **stage** basement, staff or orchestra room, store or other portion ancillary to the **stage,** immediately adjacent to and opening off the **stage,** and not separated from the rest of the **premises**.

The public mean people resorting to **places of entertainment** and recreation, whether or not they are members of a club and irrespective of payment.

Voltage definitions are as prescribed in the **IEE Regulations for Electrical Installations:**

(i) **Extra-low voltage** is a voltage not normally exceeding 50V a.c. or 120V d.c. whether between conductors or to Earth;

(ii) **Low voltage** is a voltage normally exceeding the extra-low voltage but not exceeding 1000V a.c. or 1500V d.c. between conductors or 600V a.c. or 900V d.c. between any conductor and the Earth.

Note: Chapter 13 (temporary structures) has its own additional definitions.

Chapter 1: MANAGEMENT RESPONSIBILITIES

General

1.1 It is vital that all **premises** used by **the public** are as safe as practicable from fire and that **managers** know how to protect **the public** and their staff from injury or death if a fire breaks out. To this end the **licensee** should consult the licensing authority well in advance of the proposed event to ensure that his (or her*) proposals and the fire precautions for the **premises** are adequate for the event(s) in question. The **licensee** should fully understand and comply with the conditions of the licence. It is also recommended that **managers** of non-licensed **premises** consult the fire authority for advice whenever they see a need to do so and have an evacuation plan in readiness for any emergency. The guidance which follows may therefore be regarded as being of equal importance both to **licensees** and **managers** of non-licensed **premises**.

1.2 In this chapter and throughout this guide the term **manager** is used to describe the person who has personal responsibility for the overall management of a place of entertainment. He may exercise this responsibility in his own right (for example as owner or as holder of the licence) or it may be a delegated responsibility because he has been employed for the purpose. Whatever the size of the establishment there must be no doubt where this responsibility lies. If the **manager** is for any reason absent, some other person must be given the responsibility and the authority to act in his place. A **licensee** may, of course, have specific responsibilities placed on him as part of the conditions of a licence in addition to those which might fall on him as **manager**.

The manager's responsibilities

1.3 The safety of **the public** is the responsibility of the **manager**. This means that the **manager** should be in charge of and present on the **premises** during the whole time that they are open to **the public** and kept free from work which would prevent him from being immediately available in the event of an alarm of fire.

1.4 The **manager** should also be aware that certain functions eg pop concerts and like occasions (including discos), can present additional dangers for the audience, largely from the effects of over excitement and irrational behaviour; and the noise level and flashing lights. In such circumstances, and particularly where there is a mainly younger

*The use of the words "he", "his" and "him" in this guide are simply to make the text less cumbersome to read. Many **managers/licensees** will of course be women.

audience, he should ensure that there are a sufficient number of competent and adequately trained attendants to cover an emergency situation; and a public address system which can override the performance and be heard clearly in all parts of the **premises**. The public address system should be under the direct control of the **manager**, and the system should be designed and installed to **British Standard** 5839 Part 1. It is recommended that in entertainments of this sort the **manager** should make a specific announcement to the audience at the commencement of the event (or at an appropriate occasion in the event) highlighting the whereabouts of all exits, the role of the attendants (and how they can be identified) and the procedure for evacuation in the event of an emergency. See also paragraphs 1.16, 1.17, 7.21-24, 8.2 and 8.27.

1.5 Before **the public** are admitted to any performance the **manager** is responsible for ensuring that a check is made on the following:–

(a) EXITS (to ensure that they are unlocked and available for use).
A record should be made in a log book kept for that purpose that:–

(i) all doors, gates and shutters referred to in paragraph 5.107 are locked in the open position;

(ii) all removable fastenings have been removed from other doors etc in accordance with paragraph 5.109 and placed in the **manager's** charge; and

(iii) all panic bolts have been tested and are in good working order.

(b) ESCAPE ROUTES

(i) to see that they are free of obstruction and can be safely and effectively used at all material times;

(ii) to see that **fire doors** are not wedged or propped open (although they may be held open by an **approved** automatic door release); and

(iii) to see that all exit signs are illuminated.

(c) THE **PREMISES** (to ensure that there are no obvious fire hazards).

Note: *Particular care should be taken if contractors have been working in the **premises**.*

1.6 During the performance the **manager** is responsible for ensuring that (where appropriate) the safety curtain is lowered and raised.

1.7 The **manager** is also responsible for ensuring that the following are provided and maintained:–

(a) FIRE INSTRUCTION NOTICES

(b) FIRE FIGHTING EQUIPMENT

(c) A FIRE ALARM SYSTEM

(d) ELECTRICAL AND GAS INSTALLATIONS

In order to ensure that equipment is maintained in a safe condition the **manager** should ensure that it is examined regularly by a competent person who, if not a qualified person (as described in paragraph 6.2), has the authority to engage a qualified person to carry out installations or repairs.

Note: *Any residual current device, earth leakage circuit-breaker or similar device used to connect equipment to the electrical installation of the* ***premises*** *should be checked to ensure that it is suitable for the purpose.*

1.8 The **manager** should also ensure that floor coverings, furniture and furnishings, **scenery**, **props**, curtains, drapes and hangings meet the appropriate standards. (See Chapters 4 and 8).

1.9 The **manager** is also responsible for STAFF TRAINING in respect of fire precautions and evacuation procedures. This means that he has to ensure that every member of staff is instructed and trained in the following:–

(a) the action to be taken upon discovering a fire;

(b) the action to be taken upon hearing the fire alarm;

(c) raising the alarm, including the location of the alarm call points and alarm indicator panels;

(d) the correct method of calling the fire brigade;

(e) the location and use of fire fighting equipment;

(f) knowledge of escape routes, including any stairway not in regular use;

(g) knowledge of the method of operation of any special escape door fastenings;

(h) appreciation of the importance of **fire doors** and the need to close all doors at the time of a fire and on hearing the fire alarm;

(i) the operating of all escape doors not in regular use, to ensure that they function satisfactorily; and

(j) evacuation of the **building** to an assembly point at a **place of safety** (to include reassuring **the public** and escorting them out of the **premises**). (See also Chapter 2).

It is particularly important that the **manager** ensures that where, for operational reasons, staff are transferred to another part of the **premises**, they are made aware of the **means of escape** and fire procedures of the new work area (including any additional responsibilities) if this differs from their permanent duty station.

1.10 The **manager** should keep records of staff training, to include the names of the staff attending each practice drill, and details of the training and instruction given. The following are examples of matters which may need to be included in such a record:–

(a) date of the instruction or exercise;

(b) duration;

(c) name of the person giving the instruction;

(d) name of the person(s) receiving the instruction; and

(e) the nature of the instruction, training or drill.

1.11 Practice fire drills should be held at least once every month and should be conducted by the **manager** or a competent person appointed by him. They should be attended by at least one-third of the staff including any member who has been appointed since the previous practice drill or who has not attended a practice drill within the preceding three months. Where one-off performances are given, all staff should be made aware of their duties prior to the event.

Note: The fire brigade should be called to every outbreak of fire.

Staff fire instruction notices

1.12 Printed notices should be displayed at suitable positions in the **building** stating in concise terms the essentials of the action to be taken upon discovering a fire and on hearing the alarm.

Attendants

1.13 There should be competent attendants on duty during the whole time that **the public** are on the **premises**. These attendants should have been specifically instructed as to their essential responsibilities in the event of fire or other emergency. Account should be taken of the additional responsibility caused by the attendance of disabled persons (see Chapter 2) and children (see paragraphs 1.17 to 1.19).

1.14 The primary duty of attendants is to ensure that safe conditions are maintained in the **premises** and to achieve this, they should:–

(a) ensure that no overcrowding occurs in any part of the **premises**;

(b) keep all gangways and exits clear at all times;

(c) prevent standing on seats or furniture; and

(d) be aware of any special requirements needed to ensure the safe evacuation of the audience/patrons.

1.15 Attendants should be readily identifiable to **the public** by means of some conspicuous clothing or marking system which is visible under all lighting conditions.

1.16 The number of attendants on duty on the **premises** to assist persons entering or leaving should be not less than 1 for every 250, or part of 250 persons present; and

(a) if the number of persons on the floor or tier is less than 100, there should be at least 1 additional attendant on duty on that floor or tier;

(b) if the number of persons present on any floor or tier exceeds 100 there should be at least 2 additional attendants on duty on that floor or tier.

Note: *The licensing authority may decide that because of the type of entertainment and/or the composition of the audience additional attendants should be on duty. (See paragraph 1.4).*

1.17 Where most of the audience are under the age of 16, the number of attendants on duty should be not less than one for every 100 or part of 100 within the auditorium and one attendant for every 50 or part of 50 above the lowest floor.

1.18 Places of entertainment such as leisure centres in which the disposition of facilities involves the separation of parents or guardians from their children should if possible be so arranged that the separate facilities are confined to one floor level. Where parents/guardians and their children are likely to use different floors the **building** should be provided with generous **means of escape** and the arrangement for marshalling during an evacuation of the **building** should be designed to give special confidence both to parents or guardians and children. (See also paragraph 5.9).

1.19 Where children are to be accommodated separately from their parents or guardians there should be sufficient attendants, of 16 years of age or over, present at all times to facilitate an evacuation. All attendants who have responsibility for children should have been trained as described in paragraph 1.9 and have attended at least one practice fire drill (see paragraph 1.11).

1.20 The **manager** and those in the management team whose normal duties are other than the assistance of persons entering the auditorium or entertainment area should not be included in calculating the number of attendants for the purpose of paragraphs 1.16 and 1.17. Similarly a commissionaire stationed at the entrance, or anyone else employed in an area away from the auditorium, should not be included. However, attendants who are on duty at a door leading directly into the auditorium may be counted for this purpose, even if they normally stand just outside the door.

Theatre firefighter

1.21 The licensing authority will in some circumstances require a uniformed firefighter who has practical experience in fire prevention and extinction to be present at all times when theatre **premises** are open to **the public**.

Restriction of smoking

1.22 Smoking should be strictly prohibited on **stage** and in all areas associated with the **stage**, except where it is necessary in connection with the performance. The use of naked lights will require **approval**.

1.23 In the interests of safety it is recommended that smoking should be restricted to staff rest areas only and prohibited in all parts to which **the public** have access except bars and main foyers.

Exclusion and admission of the public

1.24 All entertainments should immediately stop and **the public** should be instructed to leave the **premises**:–

(a) if there is a suspected gas leak;

(b) if the **normal lighting** fails and remains inoperative (see paragraphs 6.28 to 6.30) for more than the rated duration of the **emergency lighting** system less one hour. **The public** should not be re-admitted until the requirements of paragraphs 6.28 to 6.30 can be met;

(c) if the **normal lighting** fails, and the **emergency lighting** has a rated duration of less than one hour;

(d) if the capacity of the source of supply for the **emergency lighting** can no longer maintain that lighting for at least one hour;

(e) if the **emergency lighting** fails to comply with the level of illumination required by paragraphs 6.31 to 6.37; or

(f) in the case of small **premises**, if the **emergency lighting** fails, unless a rapidly available alternative lighting system, meeting the requirements of paragraphs 6.31 to 6.37 and, if necessary, paragraph 6.44 is brought into operation without delay and the **normal lighting** is still operating.

Notes: 1. *If the water supply fails during a performance (where this is relevant to the installed fire protection system), there should not be a need to evacuate the* ***premises*** *but no member of* ***the public*** *should be subsequently admitted until the normal supply is restored.*

2. *The term 'rated duration' in (b) and (c) above is defined as the period of time that the luminaire can continuously provide the minimum illuminance required in the emergency condition.*

Notes/Amendments

Chapter 2: ACCESS AND MEANS OF ESCAPE FOR DISABLED PEOPLE

Introduction

2.1 It is important that in a society which values the rights and amenities available to all, fire regulations should not be used to place unnecessary restrictions on the ability of disabled persons, particularly those whose mobility is impaired, to use **places of entertainment**. Although it will be for management to decide on the extent of any special provision which should be made, the assumption should be that every opportunity will be taken, so far as practicable, to enable disabled people to attend **places of entertainment** as integral members of the audience.

2.2 British Standard 5588: Part 8 - Code of practice for means of escape for disabled people, provides guidance on the measures to be incorporated into new **buildings**, or existing **buildings** which have been altered, to enable the safe evacuation of disabled people in the event of fire. It is accepted that it may not be possible to fully apply the code in existing **buildings** and in these circumstances alternative ways of meeting its objectives should then be sought.

2.3 The licensing authority should consult with representatives of disabled people, the **licensee** and the fire authority before any restriction is made regarding the admission of disabled people.

Wheelchair users and people with impaired mobility

2.4 When granting, renewing or transferring a licence, there should be agreement between the licensing authority, fire authority and the **licensee** on the number of wheelchair users who may be admitted at any one time. The fire authority should also advise the licensing authority of any additional safeguards which could be taken to permit organised parties of wheelchair users to attend a performance where numbers would exceed those normally allowed.

2.5 Whilst some existing **places of entertainment**, because of the structure and internal layout of the **premises**, will have some difficulty in accommodating disabled people in wheelchairs, in others access facilities have been provided to an increasing extent and this practice should be encouraged. It is also recommended that, where practicable, the licensing authority should encourage the **licensee** to provide wheelchair spaces in various parts of the seating area so that wheelchair users can sit with friends and have a choice of seats. Where it is proposed to make structural alterations to the **premises** the special needs of disabled people should be identified and note taken of the guidance contained in **British Standard** 5588: Part 8.

2.6 Unaccompanied disabled people who are unable to walk either aided or unaided should normally enter and leave the **premises** in a wheelchair.

2.7 Although access for wheelchair users is easier to the ground floor, the licensing authority should take into consideration means by which the wheelchair user can be admitted to other floors without prejudice to safety. It may be possible for some of the fixed seating to be removed permanently or temporarily to provide spaces for wheelchairs. This can best be arranged in the front or back rows of blocks of seating or at the ends of rows of seating. The number of seats, including wheelchair spaces, should not exceed the number **permitted**.

2.8 Where a disabled person sits in their wheelchair as part of the audience, the wheelchair should be placed in a position where it will not obstruct other people in an emergency and where a ready means of exit is available. Where a disabled person leaves their wheelchair to occupy a seat, the wheelchair should be left readily available but where it will not cause an obstruction to an exit route.

2.9 Ramps to be used by wheelchair users should conform to **British Standard** 5810.

People with impaired vision

2.10 People with impaired vision or colour perception may experience difficulty in recognising fire safety signs. Signs should therefore be sited so that they can be easily seen and are readily distinguishable. If a **manager** wishes to obtain advice about the siting of notices, this may be obtained from the Royal National Institute for the Blind or the National Federation of the Blind of the UK.

People with impaired hearing

2.11 People with impaired hearing may experience difficulty in hearing a fire alarm. However impairment of hearing does not mean that a person is completely insensitive to sound. Many people with severe impairment have sufficiently clear perception of some types of conventional audible alarm signals to require no special provision. Where this is not the case, there will in most situations be people about who can alert those with impaired hearing to the need for evacuation, and it will be reasonable to rely upon these others to provide the necessary warning.

2.12 In certain situations, such as **premises** where profoundly deaf people form the major part of the audience, alternative types of alarm signal may be necessary, eg lights or other visual signals, vibrating devices, or sound signals within carefully selected frequency bands. Technical advice on the selection of suitable devices may be obtained from the Royal National Institute for the Deaf. If an alternative alarm system is proposed, advice should be sought from the fire authority.

2.13 Induction loop systems, used in some **premises** for audio communication with people with suitable hearing aids, are not considered a satisfactory means of alerting hearing-impaired people in the event of fire. Where such systems are in normal use, however, they may be used to supplement the alarm.

Use of lifts

2.14 Disabled people may rely on a lift as a **means of escape** only if it is an evacuation lift or a firefighting lift operated under the direction and control of the **manager** using an agreed evacuation procedure. The recommendations in paragraph A3 of Appendix A to **British Standard** 5588: Part 8 should be followed.

Sources of advice

2.15 Sources of advice are:–

Access Committee for Wales/Cyngor Cymru I'r Anabl, Llys Ifor, Crescent Road, Caerphilly, Mid Glamorgan, CF8 1XL.

Centre on Environment for the Handicapped/Access Committee for England, 35 Great Smith Street, London SW1P 3BJ.

Disabled Living Foundation, 380-384 Harrow Road, London W9 2HU.

Joint Committee on Mobility for Disabled People, Woodcliffe House, 51a Cliff Road, Weston-Super-Mare, Avon BS22 9SE

National Federation of the Blind of the UK, Unity House, Smyth Street, Westgate, Wakefield, West Yorkshire, WF1 1ER.

Royal Association for Disability and Rehabilitation, 25 Mortimer Street, London W1N 8AB

Royal National Institute for the Blind, 224 Great Portland Street, London W1N 6AA

Royal National Institute for the Deaf, 105 Gower Street, London WC1E 6AH.

Scottish Council on Disability, Princes House, 5 Shandwick Place, Edinburgh, EH2 4RG.

Note: *National and local organisations representing disabled and sensory-impaired people are listed in Yellow Pages.*

Notes/Amendments

Chapter 3: CONSTRUCTION, CLASSIFICATION AND SITUATION OF BUILDINGS

Construction and classification

3.1 Since the publication of the first issue of the Home Office Manual of Safety Requirements in Theatres and Other Places of Public Entertainment in 1934, various authoritative sources have expressed differing views on the construction and classification of **buildings**. For instance Post War Building Studies No 20 grades seven classes of buildings according to their respective degree of resistance to fire; the Building Regulations 1985 Approved Document B2/3/4: Fire Spread: Table 0.1 lists nine classes of **building** in two groups on the basis of use or occupation, while the Building Standards (Scotland) Regulations 1981 define a total of 14 sub groups again on the basis of use or occupation. The existing practice in the fire service (Notes on the completion of Form FDR1) is to classify **buildings** according to their construction. *For the purpose of this guide it has however been decided to retain the classes listed in the original Manual of Safety Requirements as A, B and C* which, allied with their respective evacuation times, will enable flexibility to be exercised by the appropriate enforcing authority when determining the number of exits etc and other requirements necessary for particular **premises**. In retaining the classification a reappraisal of the interpretation of these classes has been made and a simplified version is shown in paragraphs 3.2, 3.3 and 3.4.

3.2 Class A

(a) **Buildings** constructed or substantially altered since the coming into operation of the various Building Regulations ie those for England and Wales, Scotland and Inner London;

(b) **Buildings** constructed prior to the introduction of the Building Regulations described above where evidence is available that the **building** is of not less than 60 minutes **fire-resisting construction** and the classification of surface spread of flame of walls and ceilings does not fall below Class 1 when exposed to test in accordance with **British Standard** 476: Part 7, or **Class O** as defined in A.8 of Approved Document B2/3/4 to the Building Regulations 1985. (In Scotland **Class O** means as prescribed by the Building Standards (Scotland) Regulations 1981/87.)

Maximum evacuation time: 3 minutes.

3.3 Class B

Buildings constructed prior to the coming into operation of the

Building Regulations except those which comply with paragraph 3.2 (b).

This group comprises those **buildings** loosely described as being of traditional construction with brick walls, timber floors and slate/tiled roof timber truss which form the bulk of existing **buildings** used as **places of entertainment**, ie town halls, church and community halls and many of the older theatres.

Surface spread of flame classification should not fall below Class 1 on escape routes and Class 2 elsewhere in the **building,** as defined in **British Standard** 476: Part 7 and the **building** should be of not less than 30 minutes **fire-resisting construction**.

Maximum evacuation time: $2\frac{1}{2}$ minutes.

3.4 Class C

Buildings of combustible construction (unless lined internally with fire resisting materials to afford a minimum degree of 30 minutes protection in which case they would come under Class B).

This class also includes tents, pneumatic structures and **buildings** containing considerable quantities of soft wood panelling with concealed voids.

Maximum evacuation time: 2 minutes.

It has been stated by several enforcing authorities that any reference to the Class C category should be omitted from this guide and that they themselves would not consider licensing such **buildings**, which is a point of view the Association of British Theatre Technicians also support. It would appear, however, that there still remain a large number of these **premises** and it is considered that, provided exit facilities, fire fighting equipment and other similar measures are increased pro rata to the risk, the use of this class of **premises** should continue to be allowed. (See Chapter 13 on temporary structures).

Note: *In Class A, B and C above, the "maximum evacuation time" is an arbitrary figure used to determine* ***distance of travel****, width of escape routes and the number of exits which theoretically would allow a given number of persons to escape to a* ***place of safety*** *in a specified time. (See paragraph 5.26).*

Situation

3.5 Buildings used as **places of entertainment** which form part of a modern town centre redevelopment scheme pose special problems with regard to access and frontage. Such **buildings** may be found situated:–

(a) on an elevated podium deck;

these **premises** are likely to be served by an entrance from a lower level within a covered shopping centre and would require the provision of adequate fire separation at low level in order to prevent the spread of fire and/or smoke from the shopping centre upwards to the **place of entertainment**. Where the internal **means of escape** arrangements are regarded as insufficient, provision should be made for direct access to the open air from the upper level to the podium level and from the podium level to the ground level or to a **place of safety**. In such circumstances it may be necessary for open air vehicular service roads and ramps to serve as escape routes and also as a means of access for fire appliances.

(b) enclosed within the confines of a complex;

because of the location and lack of external walls, these **premises** should be provided with an adequate number of exits, which should be served by **protected routes.** It is also necessary to ensure that there is adequate fire separation between the different occupancies in accordance with the provisions of the Building Regulations for differing purpose groups.

(c) in a piazza or precinct used exclusively by pedestrians;

these **premises** are generally situated away from service roads and because of this it may be necessary to consider whether the structure of the piazza or precinct should be strengthened or reinforced especially if access has to be provided for fire appliances. There should also be sufficient space on the piazza in order to ensure that persons making their escape from such **buildings** are able to immediately disperse to a **place of safety**.

3.6 Some existing **premises** can present special problems in respect of their situation and it will be necessary for the licensing authority to exercise the maximum flexibility in order that the **premises** may be licensed. **Premises** should however meet the following minimum criteria:–

(a) the **premises** should abut upon and have frontage on to a street or streets, or way or ways, or open space having direct access to a street, sufficient to ensure the rapid dispersal of persons from the **premises** in the event of fire, and to afford satisfactory access and working space for fire brigade appliances;

(b) the width of any street, way or open space referred to in (a) above should be not less than the aggregate width of all exits from the **premises** discharging into such street, way or open space;

(c) any private way or open space forming part of the means of egress from the **premises** should be under the control of the **licensee** although such way or open space need not be exclusive to the **premises** if existing arrangements are shown to be effective for preserving adequate **means of escape** from the **building** and

(d) any way or open space other than a public highway should:-

(i) be provided with a smooth and even surface and with adequate means of illumination; and

(ii) where considered necessary should be provided with raised kerbs or other protection for pedestrians from wheeled traffic.

Special arrangements for premises that form part of a building

3.7 Where **premises** form part of a **building**, the use and construction of the remainder of the **building** should be such that, in the event of fire, there will be no undue threat to the safety of the occupants of the **premises**. Where separation is offered by fire resisting screens and **fire doors**, this will be deemed to meet the period of fire resistance. In other cases the risks will need to be evaluated by the licensing authority.

Notes/Amendments

Chapter 4:

FIRE RESISTANCE AND SURFACES OF WALLS AND CEILINGS; FLOOR COVERINGS, FURNITURE AND FURNISHINGS

Fire resistance

4.1 When planning fire precautions and **means of escape** in **premises** it is necessary to have regard to the fire resistance of the elements of structure, eg walls, floors, etc. In the types of **premises** covered by this guide it may not always be possible to achieve the minimum standards set out in the table below. In such circumstances compensating features are required, eg a reduction in the **distance of travel** and/or the provision of automatic fire detection and/or automatic fire suppression systems.

Table A Minimum Fire Resistance (in minutes)
(Figures in brackets refer to the notes overleaf)

	Walls	Doors	Floors
Floor immediately over a basement	—	—	60 (1 & 10)
All other floors	—	—	30 (1 & 2)
Enclosing a stairway	30 (9)	30 (3,5,7 & 9)	—
Enclosing a **protected route**	30 (9)	30 (3,5,7 & 9)	—
Enclosing a lift motor room	30 (9)	30 (5,7 & 9)	—
Enclosing a lift well	30	30 (4,5 & 7)	—
Enclosing a **compartment**	30 (9)	30 (5,7 & 9)	—
In a corridor to sub-divide it	—	20 (5,7 & 9)	—
In a stairway from ground floor to basement	60	2x30 or 1x60 (5,6,7 & 9)	—
Enclosing an **area of high fire risk**	60 (8 & 9)	60 (7,8 & 9)	60 (8,10 & 11)
Enclosing a ventilation duct	30 (1)	—	30 (1 & 2)
Enclosing a dressing room etc	—	2x30 (12)	—
Enclosing a **separated stage**	60	2x30	—
Any other door required to be a **fire door**	—	30	—

NB In respect of walls and floors see the technical definition of **fire-resisting construction**.

Notes to Table A

1. *See paragraph 4.8.*

2. *This does not include for example, a gallery floor.*

3. *Except a door to a toilet containing no fire risk, provided that the toilet room is separated by **fire-resisting construction** from the remainder of the **building**, and the door is self-closing.*

4. *Except a lift well contained within a **protected stairway** enclosure (see also paragraphs 5.89 and 5.90).*

5. *An existing door may be deemed to satisfy the necessary standard of fire resistance if it can be suitably modified in accordance with the methods recommended in the Timber Research and Development Association's Wood Information Sheet, section 1 sheet 32 "Fire resisting door set by upgrading".*

6. *See paragraph 5.67.*

7. *See paragraphs 5.49 and 5.94 (d); and the technical definition of "**fire door**". Doors to corridors forming a **dead end** and those doors separating the **dead end** where it joins a main corridor (see diagram 1), should have 20 minutes integrity unless they are affording protection to a stairway.*

8. *In certain **areas of high fire risk** eg a kitchen, cloakroom, servery, etc a 30 minute door and/or service hatch or shutter may be **permitted**. During the period that **the public** are on the **premises**, and provided that the cloakroom, servery etc is attended, the door, hatch or shutter may be held in the open position.*

9. *(a) Fire resisting glass may be incorporated into the wall separating any accommodation from a corridor that is a **protected route** provided that the glass in its framework satisfies the fire resistance performance criteria for **British Standard** 476: Part 22.*

   *(b) A glazed screen with timber frames may be used to separate an escape corridor from a stairway enclosure provided that the appropriate fire resistance given in the table is maintained between the two areas. The insulation criterion of **British Standard** 476 may be waived where the glazing is limited to an area at least 1.1m above the adjoining floor level.*

   *(c) Vision panels incorporated into a door which protects an escape route should not reduce the fire resistance required for the door and should accord with the appropriate provisions of **British Standard** 6262.*

10. *See paragraph 4.2.*

11. *Other than ground floor.*

12. *Where access is allowed between the* ***stage*** *or* ***platform*** *and the dressing room, and between dressing rooms etc and the auditorium or hall, a* ***protected lobby*** *or* ***protected route*** *should be provided. The requirement will be met by having at least two 30 minute* ***fire doors*** *between the dressing room and* ***stage/platform*** *area or between the dressing room and any other public area.*

4.2 The period of fire resistance suggested in paragraph 4.1 should wherever practicable be achieved except that in a **building** where the main elements of structure offer less than a 60 minute standard, an enclosure to an **area of high fire risk** and the floor immediately over a basement should, *under no circumstances*, be less than 30 minutes fire resistance.

Surface finishes on escape routes

4.3 In the case of new **buildings** this aspect is stringently controlled by the Building Regulations but the Regulations cannot be used to control the surface finishings occasioned by redecoration long after **buildings** have been occupied. Approved Document B2/3/4 to the Building Regulations 1985 provides for the purpose group "Assembly" to conform to Table 5.4 of the Document.

4.4 Any reference to a surface being other than **Class 0** should be construed as a requirement that the wall or ceiling should be so built that a specimen constructed to the same specification, if exposed to test in accordance with **British Standard** 476: Part 7, would comply with the test criteria as to surface spread of flame specified in relation to that class. However, a wall or ceiling should be deemed to have a surface of the requisite class if it is constructed to the same specification as that of a specimen which, prior to 31 August 1973, was either proved to satisfy the relevant test criteria prescribed in clause 7 of **British Standard** 476: Part 1, or was assessed by an appropriate authority as capable of satisfying those criteria. In relation to a requirement that a surface should be of a class not lower than a specified class, **Class 0** should be regarded as the highest class, followed in descending order by Class 1, Class 2, Class 3 and Class 4.

4.5 The Building Standards (Scotland) Regulations prescribe the following standards:–

In circulation spaces and protected zones, **Class 0**.

In all other rooms, Class 1.

And certain limited concessions in the following circumstances:–

— in any room in a **building** which has an occupant capacity exceeding 10, 15 per cent of the aggregate area of the surfaces of the internal walls and ceilings may be of a class not lower than Class 3.

- the surfaces of internal walls and ceilings in any room having an occupant capacity not exceeding 10 (other than a kitchen) may be of a class not lower than Class 3.

Any reference to a surface being of **Class 0** should be construed as:–

- the material of which the wall or ceiling is constructed should be non-combustible throughout; or

- the surface material (or, if it is bonded throughout to a substrate, the surface material in conjunction with the substrate) should have a surface of Class 1 and, if tested in accordance with **British Standard** 476: Part 6, should have an index of per formance (I) not exceeding 12 and a sub-index i_1 not exceeding 6: provided that the face of any plastics material Type 1 should not be regarded as a surface of **Class 0** unless:–

- the material is bonded throughout to a substrate which is not a plastics material in a manner which will not be affected by heat or flame and the material in conjunction with the substrate satisfies the test criteria prescribed above; or

- the material satisfies the test criteria prescribed above and is used as the lining of a wall so constructed that any surface which would be exposed if the lining were not present satisfies the said test criteria and is the face of any material other than a plastics material Type 1.

Surfaces of walls and ceilings etc

4.6 Apart from small areas of wall surface (for details see Appendix A) the surface finish of walls and ceilings should correspond to a standard not lower than that indicated in the appropriate surface spread of flame classification given below:–

Class 0 In circulation spaces and escape routes

Class 1 In rooms, including auditoria and places of assembly other than small rooms

Class 3 In small rooms (not exceeding $4m^2$)

*Notes: 1. **Class 0** means the standard as prescribed by paragraph A.8 of Approved Document B2/3/4 of the Building Regulations. (In Scotland **Class 0** means as prescribed by the Building Standards (Scotland) Regulations.)*

*2. Classes 1 and 3 mean tested in accordance with **British Standard** 476: Part 7.*

*3. Where Class 1 is required, the proper application of special coatings on existing surfaces to achieve this rating may be **acceptable**, subject to suitable maintenance of the coatings.*

4. *Appendix A gives examples of materials which may be found in situ which may be **acceptable** in meeting these standards.*

5. *Subject to the concessions in paragraph 4.5 the above are **acceptable** in Scotland.*

4.7 Partitions, space dividers and other similar vertical surfaces which are provided to sub-divide a room should be not less than the class of surface required for the room in which they are situated.

Fire/smoke spread

4.8 In addition to the measures already recommended for **protected routes, areas of high fire risk**, lift wells etc (see paragraph 4.1) it will be necessary to safeguard the **means of escape** against the spread of fire or smoke and hot gases by way of service openings in the **building** structure, eg service ductwork, pipework openings, chutes and ventilation trunking. This list is not exhaustive but is intended to highlight those engineering services commonly found and which need to be considered. Safeguards normally consist of providing protective fire resisting enclosures to these services where they may pass through, or into, escape routes from other parts of the **premises**, or by effective fire/smoke stopping at the point where the **building** services penetrate the floors or walls. In the larger **premises** where extensive ventilation ductwork is present, fire dampers may be required.

4.9 Because of the need to protect the **means of escape** from the **premises** it will be necessary for a detailed examination to be made of structural separation in order to identify voids and wall cavities which may not be readily apparent. Without such an inspection and any defects being remedied, it is possible that heat and smoke could pass unrestricted through these openings thereby jeopardising the use of escape routes.

4.10 A large assembly room may not always be fully used for an event because of a reduced number attending. In such circumstances there is a risk of a fire developing in the unoccupied area and spreading to the occupied part. This is the situation which occurred at the Stardust Club in Dublin in February 1981 when a fire started in an area of seating which was not being used for the event and was thus curtained off from the rest of the ballroom. It is therefore essential that the occupied part of the **premises** and **the public** attending the event are not placed at risk from a fire developing in the unoccupied part. To ensure this, ideally the fire risk should be removed by effective separation or an automatic fire detection system installed. There is also a need to secure against unauthorised access.

Advice on floor coverings

4.11 Some floor coverings, when involved in fire, may react to produce large volumes of heat and smoke although the surface spread of flame may be relatively slow. The possibility that the floor coverings may present a hazard to **means of escape** should not be overlooked and should feature in the overall assessment of suitability of surfaces, walls and ceilings to **protected routes**. If new floor coverings are installed, they should comply with **British Standard** 5287 as conforming to the low radius of fire spread (up to 35 mm) when tested in accordance with **British Standard** 4790.

Advice on furniture and furnishings, and synthetic materials

4.12 Furniture and furnishings which are easily ignited or demonstrate rapid spread of flame characteristics present a particular fire hazard and their presence should be a factor in determining the acceptability of escape routes and in particular of **protected routes**.

Upholstered furniture

4.13 It is recommended that upholstered furniture should only contain those filling materials specified in the Furniture and Furnishings (Fire) (Safety) Regulations 1988; and that invisible parts of permanent covers comply with the provisions of the Furniture and Furnishings (Fire) (Safety) (Amendment) Regulations 1989. The furniture should satisfy as a minimum standard Ignition Source 0 (cigarette test) of **British Standard** 5852: Part 1 and Ignition Source 5 (timber crib test) of **British Standard** 5852: Part 2. However, it should be noted that **British Standard** 5852 does not test the resistance to ignition of the underside of the seating. In addition, where the cover material, or any barrier fabric, has been treated chemically to impart a degree of flame retardance, it should be subjected to the water soak test, as advocated in the Furniture and Furnishings (Fire) (Safety) Regulations 1988. The new standards should be applied when refurbishment and replacement takes place.

4.14 Application for consent for upholstered furniture should, where required, be made to the licensing authority in writing and should be accompanied by full details of the materials to be used and a certificate from an **approved** testing establishment. The licensing authority may, after consideration of specific circumstances of use, vary the level of ignition source which seating is required to meet.

4.15 All fixed and movable seating should be maintained free from tears, rips etc, which would result in the filling being exposed.

Curtains, drapes and other textile hangings

4.16 (a) All curtains or drapes hung in **places of entertainment** should be of **durably flame-retarded fabric** or **inherently flame-retarded fabric** and should conform with **British Standard**

5867: Part 2 fabric type B. Curtains or drapes should not conceal notices and should be hung so as not to trail on the floor.

(b) Attention is drawn to the Department of Environment test outlined in the "Department of the Environment Fire Retardant Specification No. 7, Ignition Standards for Curtains". Fabrics should be tested using Ignition Source 5 (timber crib test) of **British Standard** 5852: Part 2. **Licensees** may wish to consider the use of this test which subjects curtains, hangings and drapes to a more searching indication of a material's performance in a fire situation. The appropriate procedure in **British Standard** 5651 should be completed prior to this test in respect of those fabrics which have been chemically treated.

(c) Where curtains are **permitted** in front of fire exit doors they should be hung so as not to obstruct exit signs and be arranged so as not to trail on the floor and with a centre opening when in front of pairs of doors. They should only be **permitted** where attendants are present nearby to open the curtains in the event of an emergency.

4.17 The licensing authority may consider the **approval** of specific **non-durably flame-retarded fabrics**.

4.18 Temporary decorations should not be used unless **approved**.

4.19 Application for consent for all decorations, curtains and hangings should, where required, be made to the licensing authority in writing and should be accompanied by full details, including samples (not less than 1000mm x 500mm) of the material proposed to be used and, in the case of temporary decorations, a statement as to the period for which it is desired to retain the decorations.

Artificial and dried foliage

4.20 It is not possible to assess dried or artificial foliage in terms of **flame retarded fabrics** using formal laboratory test methods. It is, however, recommended that these and similar items be subjected to ignition tests using small flaming sources comparable to those used for testing drapes and curtaining, as follows:–

a suitable small flaming ignition source, the match equivalent butane flame Ignition Source 1 specified in **British Standard** 5852: Part 1, should be applied to the treated leaves, flowers, etc, of the sample for 20 seconds. Ignition is acceptable during the applica-ion of the igniting flame but, on its removal, flaming, whilst continuing locally, should not spread beyond the area first ignited.

4.21 As it has been found difficult to totally inhibit the production of flaming molten droplets or debris from the solid plastics parts of artificial foliage such as stems, the licensing authority should consider factors such as:–

(a) the location;

(b) the ease of access by **the public;** and

(c) the overall amounts of artificial foliage present.

4.22 All artificial and dried foliage used for decorative purposes in public areas should be flame retardant treated. As flame retardant treatments can be adversely affected by contact with moisture (as this can cause recrystallisation on surfaces), periodic retreatment may be required to maintain the effectiveness of the flame retardant treatment and/or to maintain an acceptable appearance. The licensing authority should therefore consider a policy of reappraisal of such treated items.

4.23 Dried flowers and grasses should not be sprayed with hair lacquer or other like substance, as such treatment will only enhance the ease of ignition and rate of fire spread.

Cellular foam

4.24 The use of cellular foam in such places as sports halls or gymnasiums may prejudice the safety of persons using other parts of the premises, if the cellular foam is involved in a fire. This is the situation which occurred at Maysfield Leisure Centre, Belfast, in 1984 where as a result of a fire involving gymnastic mats containing cellular foam 6 people died from inhaling the smoke and toxic gases. In order that the risk of fire may be reduced it is recommended that the guidance given in paragraphs 4.25 to 4.27 should be followed.

4.25 Gymnastic mats and similar equipment should ideally contain only combustion modified high resilience foam as prescribed in the Furniture and Furnishings (Fire) (Safety) Regulations 1988. The mats and equipment should also comply with the provisions of **British Standard** 1892 Part 2 when this is published. This standard is based on the Sports Council Fire Test for Mats 1987.

4.26 The contents of all new foam pits should be of combustion modified high resilience foam. Where existing pits do not contain this type of foam the contents should be replaced at the earliest opportunity. Pits should ideally be covered when not in use.

4.27 Gymnastic mats should be stored in a purpose-built store having a fire resistance of 60 minutes to **British Standard** 476: Parts 21 to 22, and where possible be ventilated to open air. When **British Standard** 1892 Part 2 is published, storage arrangements for sports/gymnastic equipment containing cellular foam should comply with the later standard.

Notes/Amendments

Chapter 5: MEANS OF ESCAPE

Introduction

5.1 This chapter deals with **means of escape** and should be regarded as guidance for use in *existing* **buildings**. **Premises** (or parts of **premises**) used as **places of entertainment** will comprise many different structural standards and interior layouts, calling for a reasonable and flexible assessment within the broad criteria of this guide. Where matters covered in this chapter have already been catered for under Building Regulations at the design stage or the **building** has been altered at some later date to Building Regulations standards, these standards should be regarded as a satisfactory alternative.

5.2 In the event of a fire breaking out in a **building** it is essential that everyone is made aware of the danger and is able to reach a **place of safety** before being overcome by smoke, toxic gas or other products of combustion.

5.3 The first awareness of a fire on the **premises** can be expected to be the result of direct observation or an announcement by a member of staff that a fire has occurred. All others in the **building** will rely upon the fire warning system. (See Chapters 1 and 9.)

5.4 On becoming aware of the need to evacuate the **premises**, persons will frequently look first to the door by which they entered but if this cannot be reached (perhaps because of the position of the fire or smoke) an understandable instinct will be to turn away from the danger and seek an alternative route to a **place of safety**. **The public** may nevertheless underestimate the risk and may be reluctant to use exits with which they are unfamiliar. The role of trained staff in ensuring that **the public** react and leave promptly should not be understated.

General principles

5.5 The basic principle, and the one on which the **means of escape** recommendations in this guide are founded, is that persons should be able to walk to safety along a clearly recognisable route by their own unaided efforts regardless of where a fire might break out in the **building**. It is however recognised that it will be difficult, if not impossible, for some people with disabilities to make their way to a **place of safety** without the assistance of others. Arrangements for disabled people should therefore be carefully considered and these are set out in Chapter 2.

5.6 Persons becoming aware of danger from fire should normally be able to turn their back on the danger and walk to a **place of safety**. This will require the provision of at least two (alternative) exits. The **distance of travel** should be limited to avoid persons being overcome by the effects of fire before reaching a **place of safety**.

5.7 The technical requirements for **means of escape** from **buildings** which are common today and the factors governing those requirements have developed through the course of this century from research, observations and experience in many countries. These factors are largely interdependent, and include evacuation time, the distance to exits, and the number, width and siting of exits. Collectively, the number and width of exits determine the 'exit capacity' which together with the siting of exits determines the maximum **permitted** number of persons – the 'occupant capacity'.

General considerations to ensure safe evacuation

5.8 To facilitate the safe and orderly evacuation of all persons on the **premises** in the event of fire the following points should be considered:-

(a) **means of escape** should be so arranged as to discharge in positions where they will not be subject to any special fire risk from the remainder of the **building** or adjoining **buildings**;

(b) the level of any part of any floor accommodating **the public** should not be more than 12m above nor more than 6m below the principal street or ground level unless:–

(i) the **building** of which such **premises** form part has **protected lobbies** provided with permanent ventilation between all floor areas, and **protected stairways**; and

(ii) the aggregate width of exits and stairways is 10 per cent in excess of the width determined by calculation in accordance with paragraph 5.26; and

(c) such other requirements as may be considered necessary in the circumstances (eg lifts for the disabled).

Accommodation of children in places of entertainment

5.9 The Inquiry into the fire at Summerland Leisure Centre in August 1973 found that parents had tended to be separated from their children since pursuits for each were located in different places and floors. Instead of going directly to exits, parents naturally tried to find their children, making their way against the flow of persons on the escape routes, thus adding to the danger. In order to minimise the risk, if children are to be accommodated separately from their parents or guardians in **places of entertainment**, the following measures are recommended:–

(a) the accommodation for children should be at or as near ground level as practicable (or the level at which the **final**

exits discharge). In no circumstances should the accommodation for children be:–

(i) on a floor above the level at which their parents or guardians are accommodated unless the route of escape is through the upper level; or

(ii) at basement level unless the children are adjacent to the accommodation for parents or guardians;

(b) the room or enclosure for children should be adjacent to an external wall and should not have fewer than 2 exits, one of which should be a **final exit**;

(c) if the room adjoins parent accommodation, the aggregate width at the exits from both areas, exclusive of the doors between the room and the parents' or guardians' accommodation, should be sufficient for the total number, ie children plus parents, guardians and other persons; and

(d) a notice should be prominently displayed where the children are deposited, saying that in the event of an emergency children will be escorted by a member of staff to a named collection point outside the building.

Note: There should also be sufficient adequately trained attendants – see paragraphs 1.16 to 1.19.

Distance of travel

5.10 The factors which have to be considered when assessing **means of escape** will vary widely from one set of **premises** to another. Accordingly, the distances suggested in the following paragraphs and Table B below should be regarded as guidelines and not as hard and fast limits. There will, for example, be many situations, and particularly in Class A **buildings**, where an increase in the **distance of travel** may be **permitted**. On the other hand, there will be instances, perhaps in a Class C **building** or structure when a reduction in the **distance of travel** will be required.

Table B Distance of travel in metres

Column 1	Column 2	Column 3	Column 4	Column 5
Building construction	(a) Escape in more than one direction. (See paragraph 5.13)		(b) Escape in one direction only. (See paragraph 5.14) (the **"dead end"**)	
	distance within a room	maximum total distance	distance within a room	maximum total distance
Class A	18m	36m	9m	18m
Class B	18m	30m	9m	15m
Class C	18m	24m	9m	12m

5.11 Recommendations relating to **means of escape** are set out in the following paragraphs to show the basic principles which can be applied. Where appropriate, **means of escape** is considered under various stages ie:–

Stage 1 Travel within rooms

Stage 2 Travel from rooms to a stairway or **final exit**

Stage 3 Travel within stairways and to **final exits**

5.12 Escape routes should be provided from any point on each floor and should be of such number, of such width and so situated as to comply with the provisions of:–

(a) paragraphs 5.10 to 5.15 in relation to the **distance of travel**;

(b) paragraphs 5.16 to 5.20 in relation to the number of persons allowed (the occupant capacity); and

(c) paragraphs 5.21 to 5.40 in relation to the exit capacity which is governed by exit widths and the number of exits.

Measuring distance of travel

5.13 Distance of travel should be measured as being the actual distance to be travelled between any point in a **building** (by way of independent and separate escape routes) and the nearest:–

(a) **final exit**; or

(b) door to a stairway which is a **protected route**; (subject to paragraph 5.45) or

(c) door to a **protected lobby**; or

(d) door to an external escape route; or

(e) door for **means of escape** in a compartment wall which leads to a suitable **means of escape**.

*Note: Where **distance of travel** is to be measured to a door in a compartment wall each of the adjoining **compartments** should have independent **means of escape**.*

Initial dead end

5.14 Where an escape route consists initially of a **dead end** and then has an alternative route to:–

(a) a **final exit**; or

(b) a door to a stairway which is a **protected route**; or

(c) a door to a **protected lobby**; or

(d) a door to an external escape route; or

(e) a door for **means of escape** in a compartment wall which leads to a suitable **means of escape**

the distance in the **dead end** should not generally exceed the distances shown in column 5 of Table B (paragraph 5.10) as appropriate to the location from which it is to be measured; and the total **distance of travel** should not normally exceed the distance shown in column 3 of Table B. Diagram 1 is an example of this point. See also paragraph 5.46.

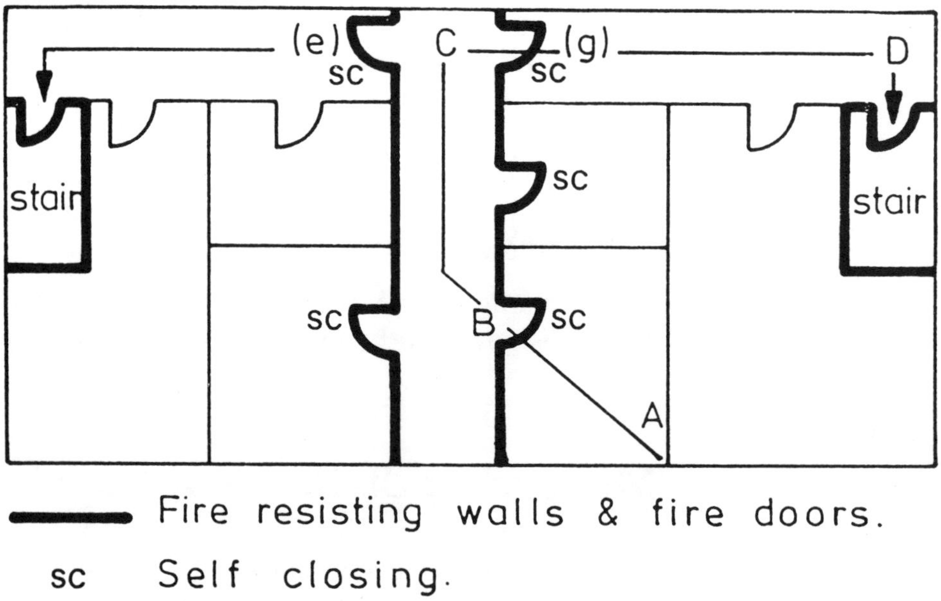

Diagram 1. Example of where dead end conditions lead to alternative escape routes.

(a) A to B should not exceed the **distance of travel** within a room (see column 4 of Table B (paragraph 5.10));

(b) A to C should not exceed **dead end distance of travel** (see column 5 of Table B);

(c) A to D should not exceed the maximum total **distance of travel** (see column 3 of Table B);

(d) Doors (e) and (g) are provided to separate the **dead end** from the alternative routes (see paragraph 5.46).

5.15 A single exit may be reasonable where persons can be readily aware of the danger and (even passing close by the source of danger) are able to reach either an exit or a point from which there are alternative exit routes without undue risk.

Occupant capacity

5.16 The occupant capacity is the number of people occupying a **building** or a part of a **building** and is an important factor in assessing **means of escape**.

5.17 In areas where fixed seating is provided the major part of the occupant capacity will be determined by the number of seats available. In other cases, however, an assessment should be made of the probable density of people within the available floor space. The figure derived from these calculations will be the 'occupant capacity'. (See Table C of paragraph 5.20).

5.18 If the maximum use is to be made of a **building**, the available exits should be of sufficient number and width to permit safe evacuation of the calculated occupant capacity. Where existing exits are not sufficient there are two courses of action open to occupiers or to the enforcing authorities. The most satisfactory arrangement is the provision of additional exit capacity by means of either more or wider exits. The other course is to limit the number of people admitted to the **premises** to that which the exits can serve, provided that the number of persons can be controlled to prevent overcrowding. Regard should also be given to the needs of disabled persons.

5.19 The calculated occupant capacity of the **premises** or any part thereof should be determined:-

(a) in areas where fixed seating is provided –

(i) if individual seats, by the number of such seats, and

(ii) if bench seats or similar continuous seating, by dividing the total width of such seating by 450mm; and

(b) in other areas (including standing areas occupied together with fixed seating) by dividing the floor area in metres ^{2} by the relevant occupant load factor given in Table C of paragraph 5.20. Toilets, stairway enclosures and similar areas are excluded; and

(c) in the case of any other room or floor not covered in Table C of paragraph 5.20, by the number of persons the room or floor is designed to hold.

5.20 The occupant load factor should not normally exceed the factors set out in Table C below:–

Table C Occupant load factors

Use of room or floor	Occupant load factor (m^2 per person)
Area for standing	0.3
Amusement arcade, assembly hall, bingo hall, club concourse, crush hall, dance hall, venue for pop concert and like occasion, queuing area	0.5
Bar	*0.3 to 0.5
Bowling alley, billiard room	9.3
Conference room, dining room, restaurant	*1.0 to 1.5
Studio (radio, film, television, recording)	1.4
Common room ie a lounge, reading room, staff room, waiting room	1.0

*depending upon the amount of seating and tables to be provided.

Note: *Where any room or floor is used or is likely to be used for a variety of purposes the occupant load factor giving the greatest occupant capacity should be applied.*

Exit capacity (related to evacuation time)

5.21 When considering the capacity of an exit route (and thus the maximum number of occupants any given exit or exit route can serve in a **building**) it should be remembered that the route will start at any occupied point in a **building**. From that point the route may follow seatways, gangways, doorways, corridors, and stairways, ending at a **final exit**. Although there are overriding factors the exit capacity of any exit or exit route is directly proportionate to the clear width at the narrowest point. However, it may be more convenient to consider the calculations in respect of the most common element of the exit route, a doorway.

5.22 In crowded conditions a file of people moving through an opening will occupy a space of about 525mm in width, and thus 525mm is known as a 'unit of exit width'. Furthermore, people will move through that space at a rate of about 40 persons per minute which gives a rate of discharge of 40 persons per minute per unit of exit width. It is questionable whether the concept of calculating exit widths only in multiples of a given dimension can be considered valid above 2 units wide. For this width of 1.05m it is accepted that people will in fact move through the exit in a more or less regular formation of two files but when the exit width is in excess of this figure observation and tests indicate that any regular pattern or formation disappears, the files become blurred at the edges and a homogeneous mass is formed.

5.23 It is against this background that there has been the development in recent years of a system or method of calculation based on increments of unit width rather than whole units. Such a method of assessing discharge rates is most useful where it is found that the proposed occupant capacity is slightly in excess of the exit discharge rate. The reason for this is that a supplementary calculation can then be made which avoids the necessity of providing an additional full unit of exit width which could be onerous, particularly in relation to a stairway. It is recommended that in **places of entertainment** the lowest incremental width to be used should be 75mm which would allow for up to 15 persons where the maximum calculated evacuation time is 2½ minutes. See Table D of paragraph 5.26.

5.24 It might be considered more expedient to relate 'exit width' directly to the width of available exits having regard to the maximum density of occupation and a calculated evacuation time which in nearly all recent guides is standardized at 2½ minutes. However, it is felt that the formulae advanced in the 'Manual of Safety Requirements in Theatres and other Places of Entertainment' issued by the Home Office in 1934 is worthy of retention and is preferable to a set table as it allows for the greatest amount of flexibility. Consequently by relating the rate of discharge per unit of exit width to the maximum **permitted** evacuation time for any **premises** the exit capacity of any exit can be determined.

5.25 The evacuation time and the exit capacity of any exit or exit route should be determined by reference to Table D of paragraph 5.26, and to the relevant class of **building** construction and by reference to paragraphs 5.29 to 5.44.

Aggregating exits

5.26 The aggregate exit capacity of any 2 or more exits should be taken to be the sum of the exit capacity of each such exit expressed in number of persons (not derived from the aggregate <u>width</u> of the exits in metres). Table D below illustrates this point (but see also paragraphs 5.29 to 5.44 on numbers and sitings of exits and note paragraph 5.33 about the need to discount the largest exit).

Table D Method by which calculations can be made to determine exit capacities

Column 1	Column 2	Column 3	Column 4	Column 5	Column 6
Building classification	Maximum calculated evacuation time	Rate of discharge per unit of exit width per minute	Exit capacity – Persons		
			Per unit of exit width 1 unit	2 units	Maximum number of persons for any exit (4 units)
Class A	3 mins	40 x 3 =	120	240	480
Class B	2½ mins	40 x 2½ =	100	200	400
Class C	2 mins	40 x 2 =	80	160	320

Summary

(i) 1 unit should not be less than 525mm = 525 mm

2 units should not be less than 525mm + 525mm = 1050mm

3 units should not be less than 525mm + 525mm + 450mm = 1500mm

4 units should not be less than 525mm + 525mm + 450mm + 450mm = 1950mm

(ii) The minimum width of a *single exit* door should not be less than 750mm.

(iii) The minimum incremental unit should not be less than 75mm after the first 2 units.

(iv) Ideally each exit should be a minimum of 2 units in width ie 525mm x 2 = 1.05m. (See Note 1).

(v) The maximum aggregate width of any exit door should not normally exceed 4 units. (See Note 2).

(vi) Revolving doors should not be considered in the calculation of available exit widths.

(vii) If turnstiles are installed they should be arranged to clear the line of exit and should not be included in the calculation of exit widths.

Notes: 1. *An exit with a width between 525mm and 1.05m is to be taken as a single unit of width .*

2. *Exit doors over 4 units in width tend to be cumbersome in use and are not considered conducive to a well ordered process of evacuation. (But see Note 3).*

3. *It should be noted that a wheelchair requires an exit width of not less than 900mm.*

Occupancy calculations

Relevant factors

One unit of exit width	525mm
Rate of discharge per minute through one unit	40 persons
Maximum permissible calculated evacuation time	
Class A building	3 minutes
Class B building	$2^1/_2$ minutes
Class C building	2 minutes
Occupant load factor	
Floor area in metres^2	

Number of persons = $\dfrac{\text{floor area in metres}^2}{\text{occupant load factor}}$

With these factors it is possible to calculate the number of units of exit width and subsequently the number and width of exits required for a given number of persons:–

Number of units of exit width

$$U = \frac{N}{40 \times T}$$

where N = Number of persons
T = Time factor in minutes
U = Number of units required

where a decimal of 0.3 or over results, the next whole number is used.

Number of exits

$$E = \frac{U}{4} + 1$$

where E = Number of exits or stairs required

Where a decimal of 0.75 or over results, the next whole number is used.

It is assumed that one exit will not be available for an evacuation. Where **premises** have a multi-purpose use then the occupant load factor should be the one for the most onerous of the uses.

Example

Note: This example demonstrates the use of rounding up (or down) as the case may be; it also brings into use the variable occupant load factors for bar areas where seating is provided.

Question:- What are the exit requirements for a single floor dance hall (Class B building)?

The dance floor area is $540m^2$
The bar area is $60m^2$ of which $30m^2$ has tables and chairs.

TO ARRIVE AT THE ANSWER YOU NEED TO COMPLETE THE FOLLOWING THREE CALCULATIONS

1. Work out the number of people that the floor area will accommodate (see para 5.20 – Table C)

(a) The dance floor will accommodate $\frac{540}{0.5} = 1080$ persons

(b) The bar will accommodate $\frac{60}{0.4} = 150$ persons

Total occupancy therefore = 1230 persons

2. Work out the number of units (U) of exit width required

The number of units (U) of exit width required is calculated as follows:–

$$U = \frac{N}{40 \text{ x } T}$$

Therefore $$U = \frac{1230}{40 \text{ x } 2.5} = 12.3 \text{ units}$$

Note: As 0.3 units attracts the rounding up rule the total is rounded up.

Total units of exit width therefore = 13

3. Work out the number of exits (E) required.

The number of exits (E) required is calculated as follows:–

$$E = \frac{U}{4} + 1$$

Therefore $$E = \frac{13}{4} + 1 = 4.25 \text{ exits.}$$

Note: As 0.25 is less than 0.75 it does not attract the rounding up rule.

Total number of exits required therefore = 4

Answer:– A minimum of 4 exits comprising not less than 13 units of exit width

Note: The mathematical calculation for exits does not take into account the need for the largest exit to be discounted (see paragraph 5.33), which means that the number of exits has to be increased in the example from 4 to 5. If the largest exit is 4 units of exit width this means that the dance hall will need to be provided with 17 units of exit width in order that 13 units are available for use at any time.

Stage 1 - Travel within rooms

Inner and access rooms (including enclosures)

5.27 Where a room is an **inner room**, ie a room only accessible through an **access room**, the distance from any point in the **inner room** through the communicating door to the exit from the **access room** should not generally exceed 9m, or 6m in **areas of high fire risk**. A situation in which an **inner room** is entered from an occupied area forming an additional **access room** should not be allowed.

Notes: 1. *An **access room** should not be an **area of high fire risk**.*

2. *A vision panel of an appropriate size should be provided in a suitable position between the **access room** and an **inner room**. This may be omitted, however, if automatic fire detection, in the form of smoke sensors, is installed in the **access room** as the occupants of the **inner room** will be aware that a fire has started. Where vision panels are installed in walls required to be fire resisting, it will be necessary to satisfy the appropriate fire resistance standard of that element. (See diagram 2 and paragraphs 9.19 to 9.21.)*

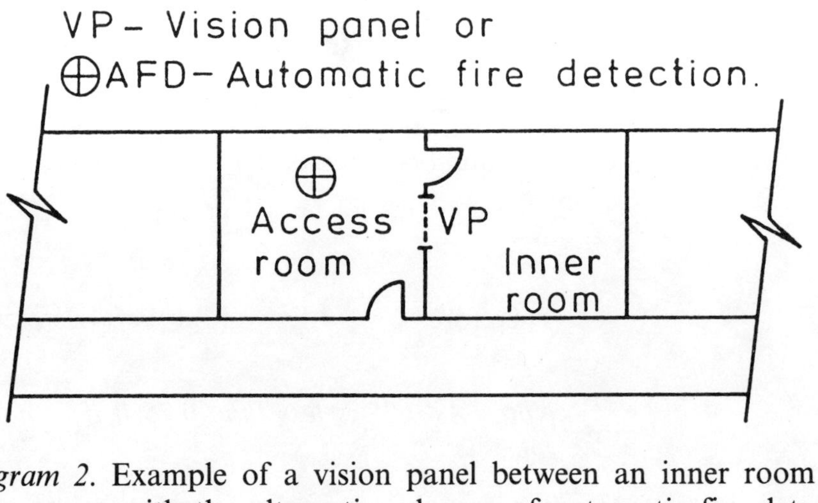

Diagram 2. Example of a vision panel between an inner room and access room with the alternative shown, of automatic fire detection in the access room.

5.28 Where the door of an **inner room** opens into an **access room** and between the rooms there is a vision panel, or if automatic fire detection is installed in the **access room**, as described in Note 2 to paragraph 5.27, the restriction on **distance of travel** recommended in column 5 of Table B (paragraph 5.10) need not apply if, from the point of exit from the **inner room,** there is escape in more than one direction through the **access room**. In such cases **distance of travel** should be in accordance with column 3 of Table B.

Number of exits

5.29 The recommendations with regard to the minimum number of exits are as follows:–

(a) in **buildings** of Class A or Class B construction in which more than 60 persons are accommodated on the ground floor, and in **buildings** of Class C construction in which more than 50 persons are so accommodated, the minimum number of exits to be provided should be two; and

(b) in **buildings** of Class A or Class B construction in which more than 60 persons are accommodated in any upper floor, circle or gallery and in **buildings** of Class C construction in which more than 45 persons are so accommodated there should be at least two exits and two stairways to ground level.

5.30 Subject to the provisions of paragraph 5.29 above at least two separate **means of escape** should be provided from every floor or tier in accordance with Table D of paragraph 5.26. The **means of escape** from each floor or tier should be remote from, and independent of, each other and should discharge into the street, way or open space.

5.31 In all cases the number of exits should be sufficient to meet the provisions in respect of **distance of travel** and (in relation to their width) exit capacity but these factors alone may not be enough to ensure a reasonable measure of safety as there is the possibility that one or more exits could be obstructed. A simple example would be that of an assembly hall which is entered through a lobby, and where a fire in the lobby would prevent use of the main entrance door (or doors). Thus in all circumstances the number of available exits from the **building** should be sufficient to meet the evacuation time limit.

5.32 The number of exits should be considered in three distinctive stages, namely exit(s) from a room, exit(s) from a floor and exit(s) from a stairway. In circumstances where both the risk and the occupant capacity is low, one exit may be reasonable (see paragraph 5.29) but normally two or more exits will be necessary (see paragraph 5.30) to allow for the possibility of obstruction (discussed in paragraph 5.31) and to avoid congestion and confusion which can often arise when a large number of people are converging at exits.

5.33 Where it is necessary to have more than one exit for **means of escape** purposes, the aggregate capacity of all the exits (see Table D of paragraph 5.26), less any one of the largest, should be capable of evacuating all of the persons in the room within the maximum evacuation time for that class of construction.

5.34 If any floor or tier is divided into two or more parts, the required exits should be provided from each part provided that, where part of the accommodation is separated from the remainder of the accommodation only by a partition allowing adequate access between the two parts of the floor or tier, such parts should not count as separate parts.

5.35 Where a room has more than one exit, both or all of which lead into an escape route which leads in one direction only to a **place of safety,** the distance from any point in the room to the nearest exit should not exceed 9m or, in **areas of high fire risk**, 6m.

Width of stairways etc

5.36 Any stairway, lobby, corridor or passageway, which forms part of the **means of escape** from the **premises**, should be of a uniform width not less than the width of the exit, or the aggregate widths of the exits, from which it leads and should be so constructed and arranged so as not to impede the free flow of traffic or the safe escape of people using it.

5.37 As a general rule stairways should be at least 1.05m wide and in all cases the aggregate capacity of stairways should be sufficient for the number of persons likely to have to use them at the time of a fire. In this connection it will be necessary to consider the possibility of one stairway being inaccessible because of fire and the aggregate width should allow for this possible reduction.

5.38 Stairways wider than 2.1m should normally be divided into sections, each separated from the adjacent section by a handrail, so that each section measured between the handrails is not less than 1.01m wide.

Measurement of width

5.39 Stairways and landings should be measured, with no account being taken of skirtings, as follows:–

(a) when enclosed on each side by walls, from the finished surface of the wall on one side to the finished surface of the wall on the other side;

(b) when constructed with a wall on one side only, from the finished surface of the wall to the outer edge of the steps and landings; and

(c) when provided with balustrades on both sides, from the outer edge of the steps or landings on one side to the outer edge of the steps or landings on the other side.

Side handrails projecting not more than 100mm and central handrails of not more than 150mm in total width may be disregarded for calculation purposes.

5.40 In the case of doorways, the width measured should be taken as being the clear unobstructed width through the doorway where the doors are open at right angles to the frame.

Siting of exits

5.41 When the minimum number of exits required to provide satisfactory **means of escape** for the **building** have been calculated, it will then be necessary to consider their location to ensure that they will fulfil their purpose in an emergency. For example, there could be in a particular **building** or auditorium a sufficient number of exits all of which could be sited at one end of the occupied area. A fire in this vicinity could conceivably prevent access being gained to any or all of the exits.

The main considerations are that:–

(a) the required exits should be as widely spaced as possible so as to allow for the occupants to turn their backs on the fire and to proceed in the opposite direction to a **place of safety**; and

(b) exits from the area involved should lead via distinct and separate exit routes to a **place of safety**. A number of exits which discharge into a common area cannot be regarded as alternatives to each other.

Note: In theatres and other ***premises*** *where the principal fire risk is on the* ***stage*** *or* ***platform****, some of the exits from each floor or tier should be arranged in positions remote from the* ***stage****.*

5.42 In a room or enclosure requiring more than one exit, the exits will be satisfactorily sited if the angle between lines defining the routes from any point in the room to the exits is not less than 45 degrees (see diagram 3(a) below), and the **distance of travel** between that point and the nearest exit is assessed to conform with the recommendations contained in column 2 of Table B (paragraph 5.10).

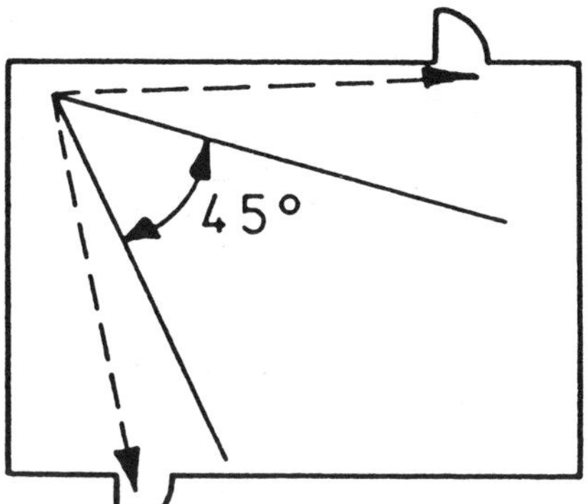

Diagram 3(a) Example of siting of exits to show 45° rule.

5.43 From any point where the route of travel is initially in one direction only, for a distance not exceeding the limitations specified in columns 4 and 5 of Table B (paragraph 5.10), to a point where there are diverging routes, the angle of divergence should be not less than 45 degrees plus $2^1/_2$ degrees for every metre travelled in one direction. (See diagrams 3(b) and 3(c).)

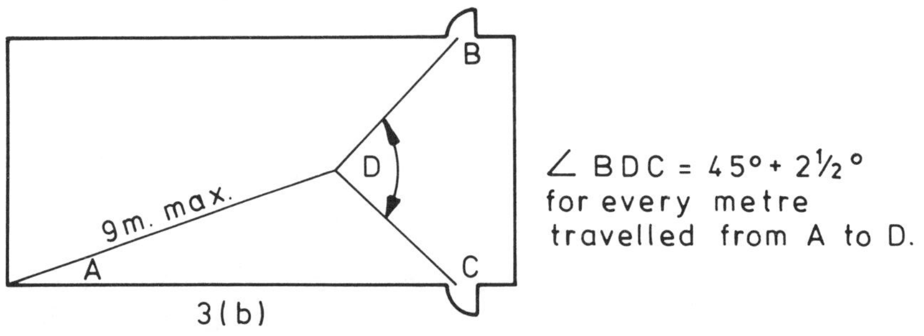

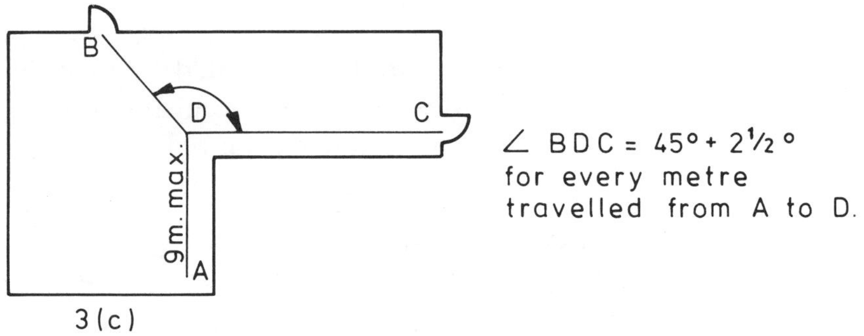

Diagrams 3 (b) and (c) Examples of sitings of exits to show the angle of divergence.

5.44 From any point in the **building** where the route or routes of travel are separated by construction having a fire resistance rating of not less than 30 minutes, there should be no restriction as to the angle of divergence.

Stage 2 – Travel from rooms to a stairway or final exit

5.45 Escape in more than one direction in Stage 2 may be deemed to be available from any point from where there are different routes leading to:–

(a) a **final exit**; or

(b) a door to a stairway which is a **protected route**; or

(c) a door to a **protected lobby**; or

(d) a door to an external escape route; or

(e) a door for **means of escape** in a compartment wall which leads to a suitable **means of escape**.

Corridors

5.46 In corridors where escape is in one direction only, the route should be a **protected route** unless it leads to a **final exit**, within the recommended **distance of travel** as shown in column 5 of Table B (paragraph 5.10). Where a corridor is required to be a **protected route** and consists initially of a **dead end** and then has alternative routes, where both sections of the corridor join, they should be separated by **fire-resisting construction**. The distance that a person should travel from any point within the **dead end** to reach that junction should not exceed the distance recommended in column 5 of Table B. (See also diagram 1, paragraph 5.14 and Note 7 to Table A of paragraph 4.1.)

5.47 Where the corridor is a **protected route**, doors to cupboards should be **fire doors** and should be marked 'Keep locked shut when not in use'.

5.48 A main corridor should not normally be less than 1.05m wide.

5.49 Corridors should be sub-divided by **fire doors** as necessary to prevent the free travel of smoke and products of combustion throughout the length of the corridor. Doors provided for the sole purpose of restricting the passage of smoke need not be **fire doors** provided that they are fitted with suitable smoke seals, are of substantial construction and self closing and double swing as necessary. Care should be taken to ensure that smoke cannot readily by-pass these doors. No undivided length of corridor should normally exceed:–

(a) Class B and C **Buildings**, 18m

(b) Class A **Buildings**, 30m

*Note: Paragraph 5.103 deals with the method of keeping a **fire door** in the open position.*

5.50 The floor of any corridor or passage should not be inclined at a gradient steeper than 1 in 12 to the horizontal.

5.51 Movement of persons up or down a group of not less than 3 steps will be so obvious to those following that they will be prepared for the change in level, but movement up or down one step is not so readily observed and may easily lead to a fall. Wherever practicable, differences of level in corridors, passages and lobbies should be overcome by the provision of inclines or ramps of gradients not exceeding 1 in 12 or steps not having less than 3 risers in any flight. Additionally all corridors and passages should be level for a distance of 1.5 metres in each direction from any steps.

Openings in floors

5.52 Where, from any point on any floor or level it may be necessary to pass near an unprotected opening in a floor through which smoke from a fire in other parts of the **building** may spread, that opening should be protected unless the occupants will be readily aware of any smoke movement and be able to turn away to an alternative route or exit.

Balconies, bridges, walkways, etc

5.53 A balcony, bridge or walkway can sometimes be used to by-pass a stairway enclosure and in some instances can form Stage 2 of an escape route. Where they form any part of the **means of escape** it will be necessary to ensure that their use at the time of a fire cannot be prejudiced by smoke and flames issuing from openings (eg windows and doors) in the external wall of the **building**. A door to or from a balcony, bridge or walkway which is intended for use as a **means of escape** should be kept unlocked. In all cases it will also be necessary to ensure that these **means of escape** can be used safely by providing, as appropriate, guard rails, hand rails, toe boards, etc. Depending on the circumstances it may be necessary to provide weather protection and lighting.

Stage 3 – Travel within stairways to a final exit

5.54 Buildings should normally be provided with two or more stairways for **means of escape**. (But see paragraph 5.29.)

5.55 Two stairways which appear to provide alternative routes in Stage 2 do not constitute alternatives for escape purposes if they both descend into one common area at ground level, unless there is **fire-resisting construction** separating the stairways at ground level so that the common area is one of negligible fire risk and access between the stairways is via two sets of self-closing **fire doors**.

5.56 Where a **building** has only one escape stairway, and surface finishes of the stairway do not conform with the recommendations in paragraphs 4.6 and 4.7, an alternative escape route should be provided.

5.57 In **buildings** where more than one stairway is necessary for escape purposes and the surface finishes of one or more stairway(s) do not conform with the recommendations in paragraphs 4.6 and 4.7, this may nevertheless be **acceptable** if there are compensating factors eg automatic fire detection or a lobby approach to the stairway(s) in question.

Enclosure of stairways

5.58 All escape stairways should be separated from the remainder of the **building** by **fire-resisting construction** and **fire doors** so as to form a stairway enclosure.

5.59 Where an escape stairway is enclosed as recommended in paragraph 5.58 with a **final exit** from the enclosure and the only doors into the stairway are:–

(a) from toilets containing no fire risk – see Note 3 to paragraph 4.1;

(b) from **protected lobbies**;

(c) from corridors;

(d) from lift wells with no openings other than those to the stairway enclosure; or

(e) a **final exit**

the stairway should be regarded as a **protected route**.

Note: *Where doors other than those listed above (eg doors from rooms or cupboards) open into a stairway, or there is no **final exit** from the stairway, it will be necessary to determine whether the stairway can be regarded as a **protected route**,*

5.60 Stairway enclosures should preferably lead direct to a **final exit**. Where there is only one stairway from the upper floor(s) of a **building** and a **final exit** cannot be provided from the stairway enclosure either:–

(a) 2 exits should be provided from the stairway enclosure each giving access to **final exits** by way of routes separated from each other by **fire-resisting construction**; or

(b) a **protected route** should be provided from the foot of the stairway enclosure leading to a **final exit**.

5.61 Where an escape stairway (including an external stairway conforming to paragraph 5.78) can be considered to be a **protected route** it will not be necessary to have regard to **distance of travel** in Stage 3. Where this is not the case the Stage 3 section of the escape route should be regarded as forming part of the total **distance of travel** (see columns 3 and 5 of Table B (paragraph 5.10)).

5.62 Where an escape stairway is a **protected route** with a **final exit** to the open air, and the stairway only serves a balcony or tier, there may not be a need for the door at the head of a stairway to be fire resistant.

5.63 Where a main stairway does not comply with the **final exit** provision of paragraph 5.59 because it discharges through the main entrance lobby (foyer) of the **premises** it may be counted as an escape stairway provided that:–

(a) there is direct and unobstructed access from the point at which the flight or flights terminate to a **final exit**;

(b) the foyer and remainder of the stairway is separated from all adjacent accommodation by **fire-resisting construction** and self-closing **fire doors**;

(c) the stairway contains nothing which is readily combustible or likely to contribute to rapid ignition or development of fire; and

(d) there remain a sufficient number of alternative escape routes from all areas served by the main stairway for the occupant capacity.

5.64 Except for doors connecting directly with the auditorium, doors in stairway enclosures should only communicate with circulation spaces, corridors or lobby approaches to the stairway.

Protected lobby approach to single stairway building

5.65 With the exception of **buildings** of not more than a ground and first floor, in a **building** with a single stairway the access to the stairway from any floor, other than the top floor or any room (other than a toilet containing no fire risk) should be through a **protected lobby** or a corridor which is a **protected route**. (See diagrams 4(a) and 4(b)).

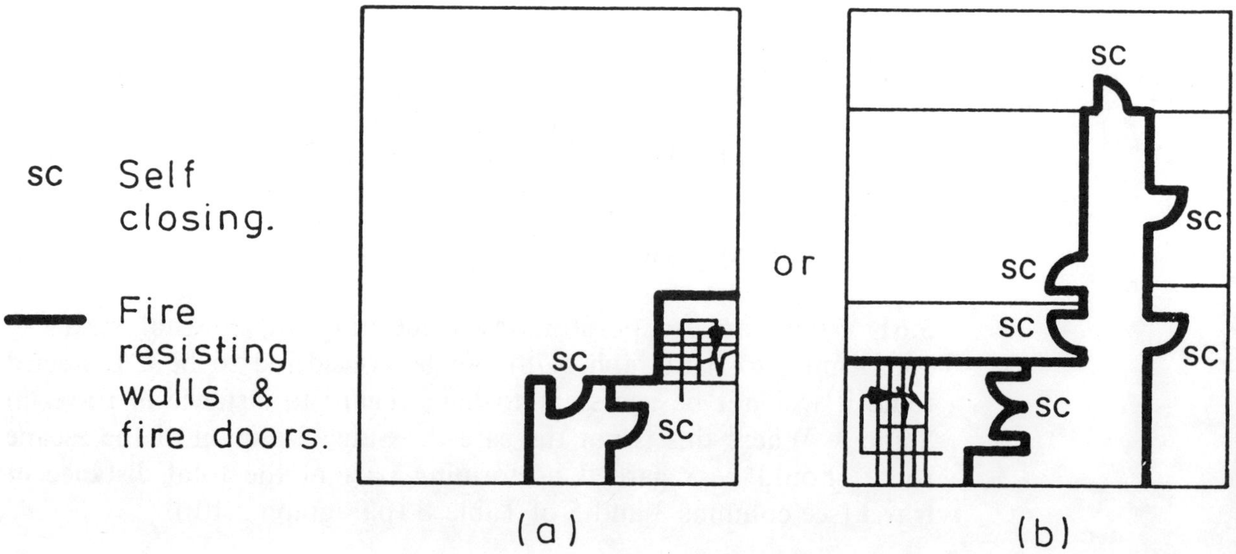

Diagrams 4(a) and (b) Examples of a single stairway separated from floor area by a protected lobby or corridor which is a protected route.

Stairways to basements

5.66 If the **distance of travel** from any point in the basement exceeds that given in column 5 of Table B of paragraph 5.10 the basement should have at least two stairways or other escape routes to a **place of safety** at ground level.

5.67 Preferably a stairway serving upper floors should not extend to the basement, and wherever possible all stairways to basements should be entered at ground level from open air, and in such positions that smoke from any fire will not obstruct any exit serving the upper floors to the **building**. Where the stairway links a basement with the ground floor the basement should be separated from the ground floor either by a **protected lobby** or by two 30 minute **fire doors**, (one at basement and one at ground floor level) or, if this cannot be achieved, by one 60 minute **fire door**. In cases where a stairway providing access for technicians links a **stage** basement through a protected **stage** to fly and lighting galleries, exemption to the above may be afforded for that stairway.

Accommodation stairways

5.68 Accommodation stairways not required for escape purposes need not be separated from the remainder of the **building** provided that:–

(a) the **means of escape** are sufficient without relying on the **accommodation stairway**;

(b) no escape route from a **dead end** of an upper floor passes through an **accommodation stairway** (see also paragraph 4.8 in relation to other floor openings);

(c) the stairways do not pass from one **compartment** to another;

(d) no such stairway serves more than two floors; and

(e) neither of the floors is connected by an open stairway to a third floor.

Headroom of stairways, lobbies etc

5.69 The clear headroom of every stairway, lobby, corridor or passage should be not less than 2m measured vertically from the pitch line of the nosings of steps and from the floor in all other cases. There should be no projections from any wall or ceiling below a clear height of 2m of any stairway, lobby, corridor or passage which may impede the free flow of persons using it.

Stairways and landings

5.70 A stairway should have:–

(a) no fewer than three risers nor more than sixteen risers in any flight;

(b) no more than two consecutive flights without a turn;

(c) a turn between consecutive flights where any flight has twelve risers or more;

(d) no riser within the opening swing of any door; and

(e) a landing at the top and bottom, and between consecutive flights.

The length of the landing should be not less than the width of the stairway of which it forms part.

5.71 The rise of any step should be uniform throughout its length, and should be the same as that of every other step in the same flight.

Spiral or helical stairways

5.72 The going of a step should be not less than 280mm and all consecutive tapered treads should have:–

(a) the same minimum going of 280mm, measured at a point 270mm from the wide end of the tread;

(b) the same rate of going and taper;

(c) their narrow end at the same side of the flight; and

(d) an angle of taper of not more than 15 degrees.

Lighting conditions for stair treads

5.73 Where lighting conditions render it desirable the nosings of all stair treads should be made conspicuous by painting or other effective means.

Escalators

5.74 Escalators are not normally allowed as a **means of escape** although in certain **premises** it may be reasonable to regard them as similar to **accommodation stairways** (see paragraph 5.68).

5.75 Escalators not within stairway enclosures may need to be separated from the remainder of the **building** by **fire-resisting construction** and by **fire doors**.

5.76 To avoid a situation where persons are carried towards a fire it is important that arrangements should be made to stop escalators at the outset of an emergency.

General

Ventilation of a stairway

5.77 Wherever practicable, there should be provision for ventilating stairways in the event of fire, particularly if the stairway enclosure is not adjacent to an external wall which has openable windows, and the stairway continues uninterrupted to the top of the **building**. The minimum area of permanent or openable venting normally regarded as satisfactory is considered to be not less than $1m^2$ or 5% of the internal area of the stairway enclosure, whichever is the greater.

External stairways

5.78 Where an external escape stairway is provided it should be a **protected route** and it will be necessary to ensure that the use of it at the time of a fire cannot be prejudiced by smoke and flames issuing from openings, (eg windows and doors) in the external wall of the **building** below and adjacent to the stairway. It will also be necessary to provide lighting (see paragraphs 6.28 and 6.31) and consider the protection of the stairway from the weather.

Ramps

5.79 Where a ramp is provided it should:–

(a) have an easy gradient and in no case should it be steeper than 1 in 12;

(b) be provided with guard rails, where it is more than 600mm above the ground, which should be at a height between 150mm and 300mm above the ramp;

(c) be provided with handrails which should give a firm support and be between 840mm and 1m above the surface of the ramp;

(d) have a width not less than that of the exit it serves;

(e) normally be divided into sections, each separated from the adjacent section by a handrail, so that each section measured between the handrails is not less than 1.01m wide; and

(f) be provided with non slip surfaces which will be particularly necessary where a ramp is exposed to the weather.

Note: Ramps for wheelchair users should conform to British Standard 5810. In particular they should be provided with level resting space landings every 10 metres and should have raised safety edges and handrails.

Handrails

5.80 A continuous handrail should be fixed on each side of all stairways, steps, landings and ramps at a height of not less than 840mm or more than 1m measured vertically from the pitch line of the nosings of steps and from the floor of landings, except that a second handrail need not be provided to such stairs and ramps or flights of steps which are not more than 1.01m wide. The handrail should not normally project more than 100mm and the ends should be turned for safety. (See paragraph 5.84).

5.81 All stepped or sloping gangways adjacent to walls should be provided with handrails affixed to the walls.

Protective barriers

5.82 A protective barrier should be provided in each of the following positions:–

(a) at each side of any stairway or ramp; and

(b) at the perimeter of any landing or floor, balcony or any other part of the **premises** where such provision is necessary to ensure the safety of persons using the **building**.

5.83 The minimum height of a protective barrier should be as follows:–

(a) barrier guarding a balcony which is immediately in front of fixed seating – 790mm;

(b) barrier guarding a flight – 900mm; and

(c) any barrier not described in paragraphs (a) or (b) above – 1.1m.

5.84 A handrail may form the top of a protective barrier provided that the appropriate height is achieved.

5.85 Any new protective barrier should be designed in accordance with **British Standard** 6180.

Barriers across escape routes

5.86 Rigid barriers should not be provided in or across any gangway, exitway or route of escape from the **premises**. Any rope barrier which may be provided should be fitted with automatic catches or slip connections and should be arranged so as not to trail on the floor when parted. The fittings should not project into the gangway or exitway.

Floors and tiers

5.87 No floor or tier whether temporary or permanent should be constructed with a slope of more than 35 degrees to the horizontal.

5.88 The floors of gangways, lobbies, corridors and passageways and the treads of steps and stairways, should have non-slip and even surfaces.

Lifts and hoists

5.89 Lifts and hoists are not normally considered satisfactory for **means of escape**. Exceptionally, however, when it is necessary to consider **means of escape** for persons in the **premises** who may be disabled, the use of a lift(s) should not be ruled out provided that the appropriate provisions of **British Standard** 5588: Part 8 are complied with.

5.90 Unless a lift is situated within a stairway enclosure it should be contained within a lift well enclosure of **fire-resisting construction** in which the access doors are **fire doors**. Existing sliding doors to lift shaft openings are sometimes ill fitting in their slides and frames and offer a poor barrier to the passage of smoke. In such cases where the opening discharges into a corridor which is a **dead end**, a **protected lobby** should be provided at the entrance to the lift. A person should not have to pass through the lobby to reach the continuing route of escape.

5.91 A lift motor situated at the foot of a lift well which is within the enclosure of a stairway which is a **protected route** and forms the only escape route from a **building** (or part of a **building**) should be housed in a **compartment** separated from the lift well by **fire-resisting construction.** Any opening necessary in the separation between the **compartment** and the well for the operation of the lift should be as small as possible.

5.92 Where practicable, a lift well should have a permanent ventilation opening at the top equal to not less than $0.1m^2$ for each lift in the lift well enclosure.

Roof exits

5.93 Persons escaping from a fire, other than those escaping from a level below ground, should not normally have to ascend to a higher level to reach a **place of safety**. However, in exceptional circumstances where it is not practicable to provide **means of escape** without placing reliance on an upwards escape route, roof exits may be **acceptable** provided that the following criteria are met:–

(a) roof exits should not normally be used by **the public**;

(b) the roof should be flat, adequately defined and the surface should be of a safe non-slip character and the route guarded with protective barriers and provided with lighting (see paragraphs 6.28 and 6.31);

(c) the escape route across the roof and its supporting structure should be constructed as a fire resisting floor;

(d) if the escape route is in one direction only any ventilation outlet or other extract system and any door roof lights, or windows that are not fire resisting should not be sited within 3m of the route;

(e) persons should not have to ascend more than one level (ie from one floor to the next or from a top floor to roof level);

(f) the exit should be in, or should lead directly to, a **place of safety**;

(g) the higher level to which persons ascend should not be to an **area of high fire risk**;

(h) the route by which the higher level is gained should be a stairway;

(i) if the stairway serves any lower level(s) also, the upward escape route should be separated from the remainder of the stairway by **fire-resisting construction** and by **fire doors**.

Doors on escape routes

5.94 A door used for **means of escape** should normally open in the direction of travel. The door should also:–

(a) be hung so that when open it does not obstruct any escape route (see diagram 5);

(b) open through not less than 90 degrees;

(c) be provided with a vision panel if it is hung to swing both ways; and

(d) if it is a **fire door**, protecting an escape route, be fitted with smoke seals.

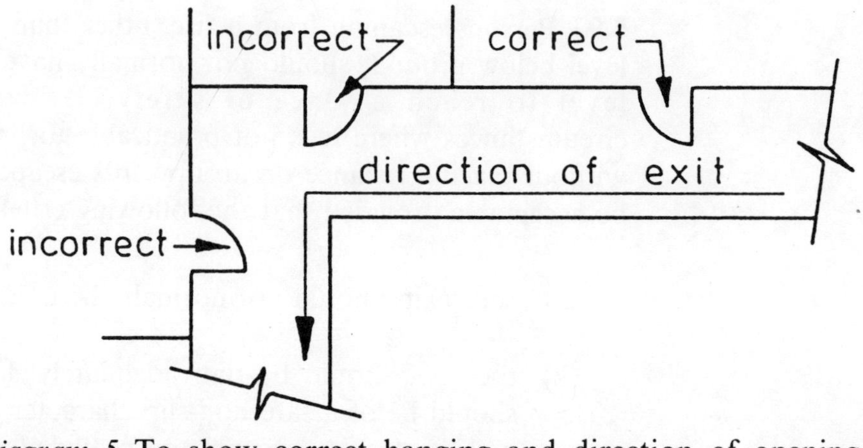

Diagram 5 To show correct hanging and direction of opening of doors into an escape route.

5.95 Any door or gate which opens into a corridor, stairway, passageway or gangway should, where possible, be recessed. If this is not possible it should be hung so as to be closed by a stream of persons passing through such a corridor, stairway, passageway or gangway in the direction of the exit. Any such door or gate should be rendered self-closing, and should be arranged so as not to interfere with exit or other doors. Consideration should be given for such a door or gate to be provided with a vision panel.

Note: Where, for convenience of working, a door not providing fire separation is required to be held open into an exitway, it should be fitted with a holding device which will allow it to be shut readily by gentle pressure on the door itself.

5.96 A door or gate should not open immediately onto a step or steps, and a landing equal to the width of the doors and not less than 1m in length should be provided between the door or gate and the step or steps. Where a door or gate is at the head or foot of steps, a landing of not less than 1m should be provided between the door or gate and the top step or the bottom step as the case may be.

5.97 Thresholds or sills of all doors should not normally protrude above the surface of the floor. Where this is a practical necessity for weather protection they should offer the minimum obstruction.

Doors and openings other than exits

5.98 Any door or opening other than an exit, which leads to parts of the **premises** to which **the public** have access should have a conspicuous notice indicating the use of such parts. Any door not to be used by **the public** should be similarly indicated or should be marked 'Private'. Such doors and openings should *not* resemble exits and the notices should *not* be specially illuminated. Notices bearing the words 'No exit' or 'No way out' should be prohibited.

Exit doors

5.99 There have been many tragic disasters, where persons have died as a result of fire, which can mainly be attributed to the fact that exit doors were either locked, obscured or otherwise obstructed. It is therefore a cardinal principle in relation to **means of escape** that all gates and doors should be able to be opened easily and immediately by persons leaving the **building** or making their way to a **place of safety**.

5.100 All the exit doors provided should lead directly to a **final exit** or a **protected route**.

5.101 The total width of all **final exit** doors, or other means of exit from a lobby/foyer, should be not less than the total width of all exitways or other means of exit leading to them.

Self-closing devices

5.102 All **fire doors**, except those to cupboards and service ducts should be fitted with self-closing devices to ensure the positive latching of the door. Rising butt hinges are not generally considered to be satisfactory. **Fire doors** to cupboards, service ducts and any vertical shafts linking floors should be kept locked shut when not in use (see paragraphs 5.121 and 5.122).

Automatic door releases

5.103 No means should be provided for holding a self-closing **fire door** in an open position other than by an automatic door release which will effectively cause the door to close when activated by a signal from the electrical fire alarm, which incorporates smoke detectors. The release should also be provided with a ready means of manual operation from a position at the door. Automatic door releases should only be fitted to a door which cannot be kept closed because of the problems presented by its frequent use in the day-to-day running of the **premises**. Normally no such device should be fitted to doors forming part of a stairway enclosure.

Note: *Doors which are held open in the manner referred to in paragraph 5.103 should be closed after any performance. This may be achieved by the installation of a remote switch.*

Fastenings on doors

5.104 Any exit door or gate which is required to be kept open whilst **the public** are on the **premises** should be provided with **approved** fastenings so that the door or gate is kept open in such a way that it will not cause an obstruction.

5.105 Doors used for **means of escape** should be kept unlocked at all times when people are in the **building** and in no case should a door be so fastened that it cannot easily and immediately be opened from the inside without the use of a key. Where the door has to be kept fastened while persons are in the **building**, the fastening should only be by means of a panic latch or panic bolt (or such fastenings as may be **approved**) which ensures that the door can be readily opened by pressure applied by persons within.

5.106 Panic bolts or latches should be without sharp projections and should be of such pattern that normal horizontal pressure on the push-bar will open the door. The push-bar should be placed at a height within the range of 900mm-1.1m from the floor. The words 'PUSH BAR TO OPEN' should be displayed conspicuously in block letters not less than 20mm, and preferably 50mm in height immediately above or below the push-bar. Panic bolts should be of a strong type and the working parts should be properly maintained. Panic bolts and panic latches fitted after the publication of this guide should comply with **British Standard** 5725: Part 1.

5.107 All doors which for structural reasons cannot be hung to open in the direction of the exit and all collapsible, sliding or roller gates, shutters or doors across exitways should be locked with a removable key, in the fully open position, at all times when **the public** are present. During such times the key should be removed and kept secure from unauthorized persons (eg in the **manager's** office).

5.108 Where the doors, gates, or shutters are required by paragraph 5.107 to be locked open, a notice or notices clearly visible on both sides of the door, gate or shutter (whether open or closed) should be prominently displayed. The notice should bear the words 'This door (gate, shutter etc) to be secured open when the premises are occupied'.

5.109 Removable security fastenings such as chains, bars, padlocks etc should be removed from all doors, gates, shutters etc on exit routes at all material times. Such fastenings should be placed on numbered hooks in an **approved** position not accessible to unauthorized persons (eg in the **manager's** office). There should be only one such numbered hook for each fastening. No removable fastening should be replaced until **the public** have left the **premises** and sufficient exits should remain available for staff who may be present.

Sliding doors

5.110 These are not normally suitable on escape routes unless they are for the sole use of staff. Where provided, a notice with the words 'Slide to open' and with an arrow pointing in the appropriate direction should be permanently displayed at about eye-level on the face(s) of all sliding doors which afford an exit or are situated across an escape route.

Exit and directional signs

5.111 People will usually attempt to leave **premises** by the same way that they entered. As these entrances in **places of entertainment** will of necessity be restricted it follows that the exit facilities provided for the **premises** will not be fully utilized. For this reason all available exits should be clearly indicated so that **the public** are aware that there are ways to leave the **building** other than by the doors which they used to gain admission. In addition, the provision of well sign-posted exits in full view of the audience will give a feeling of security in an emergency situation.

5.112 Fire safety signs, notices and graphic symbols should conform with **British Standard** 5499: Part 1, and with **British Standard** 2560 for internally illuminated exit signs. **Self-luminous fire safety signs** should also comply with the requirements of **British Standard** 5499: Part 2. Existing signs and notices which do not meet the **British Standard** need not be replaced immediately if they are effectively fulfilling their purpose. They should however be examined and replaced if they are found to be inadequate.

5.113 Self-luminous fire safety signs will not generally be allowed where the viewing distance exceeds 24m.

5.114 Maintained internally illuminated exit signs will be required where the lighting may be dimmed or extinguished.

5.115 Exit signs should be illuminated whenever **the public** are present.

5.116 A notice bearing the word 'EXIT' in plain block lettering not less than 125mm high should be placed over any door or opening leading from the place of assembly into an exit. The notice should be positioned, wherever possible, between 2m and 2.5m above the floor. No other notice should be incorporated in, or form part of, an 'EXIT' notice other than a directional arrow.

5.117 Any exit which is not a normal route of travel from the **premises** should be indicated by a notice bearing the words 'Fire exit' in conspicuous lettering of appropriate size. The notice should, wherever possible, be displayed immediately above the exit opening. Where this is not possible a position should be chosen where the notice can be seen and it is least likely to be obstructed.

5.118 Where an exit cannot be seen or where persons escaping from the **premises** may be in doubt as to the location of the exit, 'Fire exit' signs, to include a directional arrow, should be provided at suitable points along an escape route. Such notices should be fixed in conspicuous positions and wherever possible should be positioned between 2m and 2.5m above the floor level.

Notices on doors, shutters and gates

5.119 All fire notices should be in conspicuous lettering and of appropriate size.

5.120 A notice with the words 'Push bar to open' should be permanently displayed immediately above the push-bar on all doors fitted with a panic bolt or panic latch.

5.121 A notice with the words 'Fire door - Keep shut' should be permanently displayed at about eye level on both faces of all **fire doors** except those to cupboards. **Fire doors** which are normally open but which close automatically on the operation of fire detectors should bear the words 'Automatic fire door-Keep clear'.

5.122 A notice with the words 'Fire door - Keep locked shut' should be permanently displayed on the outside face of all **fire doors** not required to be self-closing eg cupboards.

5.123 A notice with the words 'Fire escape - Keep clear' should be permanently displayed at about eye level on the external face of all doors which are provided solely as a **means of escape** in case of fire and which, because they are not normally used, are liable to be obstructed.

Mirrors

5.124 Any mirrors situated on escape routes should be so sited that persons escaping from fire will not be confused by any reflected image of the route they are using, or be misled as to the direction they should take to reach fire exits.

Ventilation systems

5.125 Where ventilation systems might assist the spread of flames, smoke and hot gases from a fire it will be necessary to take steps to safeguard the **means of escape** against this hazard. (See also paragraph 4.8.)

Notes/Amendments

Chapter 6: **ELECTRICITY, GAS AND OIL**

Electrical, gas and oil equipment not to give rise to danger

6.1 All electrical, gas and oil-firing equipment should be designed, constructed, installed, protected, commissioned, tested, used, operated and maintained so as to prevent danger.

6.2 Electrical installations should be installed, tested and maintained in accordance with the provisions of the **IEE Regulations for Electrical Installations**. Work on electrical installations and appliances should only be carried out by competent persons such as the manufacturer's appointed agent, the electricity supplier, a professionally qualified electrical engineer, a member of the Electrical Contractors' Association, a member of the Electrical Contractors' Association of Scotland, a contractor approved by the National Inspection Council for Electrical Installation Contracting, or a qualified person acting on behalf of one of the above.

6.3 Gas installations in any **premises** supplied with gas through pipes from outside their boundary are subject to the Gas Safety (Installation and Use) Regulations 1984. These place a duty upon suppliers of gas, installers and users. Improperly installed or maintained gas fittings and appliances can result in fires, explosions, or the release of toxic products of combustion. Work on appliances or fittings should only be carried out by competent persons such as the gas supplier or an installer registered with the Confederation for the Registration of Gas Installers (CORGI). Gas installation should not be undertaken except by an installer who has passed an appropriate full training course or, in the case of an experienced installer, has passed an appropriate assessment test. Practical guidance on standards of training in safe gas installation is given in the Health and Safety Executive (HSE) Approved Code of Practice 'Standards of training in safe gas installation'. Whilst the Gas Safety (Installation and Use) Regulations 1984 do not apply to premises which have their own gas supply (ie where the supply is piped from liquefied petroleum gas (LPG) cylinders or a tank within their boundary), the Regulations may still be used as a guide to safe practice wherever it is reasonably practicable to do so.

6.4 Oil fired boiler installations and oil storage facilities should comply with the provisions of **British Standard** 5410: Part 1 or Part 2 as appropriate for the installation, commissioning, testing and maintenance.

Gas controls

6.5 The means of shutting-off the supply of gas to any appliance and/or to the whole of the gas installation within a **building** should be readily accessible and positioned so that access is unlikely to be impeded by **the public** in the event of an emergency. Regulations 8 and 14 of the Gas Safety (Installation and Use) Regulations 1984 are particularly relevant to emergency controls.

Electrical safety

6.6 Where the socket outlets from the electrical installation may be used for the connection of lighting, video or sound amplification equipment, display models and other portable equipment, the licensing authority may consider that each socket outlet, particularly in the **stage area** (except for socket outlets which are controlled by a properly installed **stage** lighting installation), should be protected by an integral residual current device (RCD) (earth leakage circuit-breaker) having a rated tripping current not exceeding 30 milliamps. RCDs should comply with the provisions of **British Standard** 4293.

6.7 Socket outlets in a **stage area** and their associated wiring should be regularly tested and inspected by one of the persons specified in paragraph 6.2 and a certificate showing the results of the inspection and test should be made available to the licensing authority. The licensing authority may also wish a certificate of electrical safety to be provided for each item of portable electric and electronic equipment as a condition of the licence.

6.8 Where RCDs are fitted to protect socket outlets their correct operation should be tested by operating the test button before use. If the device does not switch off the supply, an electrical contractor should be consulted and the use of the socket outlet prohibited until the defect has been rectified.

6.9 RCDs protecting circuits other than socket outlets should be tested at regular intervals not exceeding 6 months.

Notes: 1. *The use of RCDs reduces the risk of shock and fire associated with permanently plugged-in and portable equiment, but cannot eliminate all electrical hazards. Other safety precautions should not be relaxed.*

2. *RCDs as described above are not required on **normal lighting** circuits, as a fault in any part of a circuit would cause all lights on that circuit to be extinguished.*

Protection and size, strength and capacity of electrical equipment

6.10 All electrical equipment should be:-

(a) of adequate size, strength and capacity, so as to prevent danger when put to use; and

(b) suitably placed, safeguarded and protected, so as to prevent danger arising from any fault or overload.

6.11 There should be sufficient, correctly installed and rated, electrical power sockets within the **premises**, so as to preclude the use of long trailing flexes to portable electrical appliances and/or multi-socket adaptors. Where equipment is suspended by ropes, cables or other combustible materials, safety chains or other non-combustible means of support should be provided.

Electrical requirements

6.12 The electrical installation of the **premises** should be in accordance with the **IEE Regulations for Electrical Installations**, current at the time of installation, and all materials or equipment for which there is a **British Standard** or Specification should conform to the appropriate specification or code of practice, or be constructed to an equivalent standard.

6.13 In the case of new installations, a diagram of the main electrical circuits should be fixed in a position easily accessible to the technical staff.

6.14 The main circuit switch gear and fuses should be situated in a place set apart, the door of which should be kept shut and locked if the enclosure communicates with any part of the **premises** to which **the public** have access.

6.15 All main switches or circuit-breakers should be labelled to indicate the circuits they control, and the fuse ways of the sub-circuits of distribution boards should be clearly labelled.

6.16 The polarity of every circuit should be correct, in particular where single pole switches are used to ensure that the live pole is switched.

6.17 Local switches for the control of lighting in parts of the **premises** accessible to **the public** should be avoided wherever practicable, except for the provision of switches to control house and cleaners' lighting in an auditorium. In such cases, the switches should be so placed that they are not readily accessible to **the public** and should be either key-operated or otherwise enclosed to prevent unauthorised interference.

6.18 Electric signs and advertising devices on the outside of the **premises** should be controlled by switches or circuit-breakers to isolate these from the electrical supply. The switches/circuit-breakers should be situated in a position readily accessible to the staff.

Electrical wiring - general

6.19 Systems of wiring used for the **low voltage** circuits of the permanent electrical installations should have all cables and conductors enclosed throughout in protective covering of material of adequate strength to resist mechanical damage, eg armoured cables, mineral insulated sheathed cables, insulated cables protected by screwed metal conduit or trunking complying with **British Standard** 4678: Part 1 or Part 4, or rigid PVC conduit complying with **British Standard** 4607, as amended by **British Standard** 6099: Part 2.

6.20 PVC insulated and sheathed cables complying with the requirements of **British Standard** 6004 may be used for staff warning and oral announcement systems when the voltage does not exceed **extra-low voltage** and the cables are routed through protected areas of low fire risk. Where this is not possible, a system of wiring as detailed in paragraph 6.19 above should be used. Where such systems form part of any emergency signalling or alarm system the wiring should comply with the provisions of **British Standard** 5839: Part 1.

6.21 Where cables, conduit or ducts pass through walls, ceilings or partitions, the surrounding hole should be filled with fire-stopping material to prevent the passage of fire or smoke.

6.22 Where the installed batteries give off a flammable gas when being charged, the conductors in the battery room from the batteries to the control equipment should not be enclosed in conduit or ducting. Where the conductors from the batteries pass from the battery room to another room or enclosure, provision should be made to prevent flammable gases passing from the battery room to the other area via the cable run. (See Appendix B.)

6.23 Cables to be installed outdoors on walls and the like should incorporate a sheath and/or armour suitably resistant to any mechanical damage likely to occur, or be contained in a conduit system or other enclosure affording adequate protection against such damage.

6.24 No wiring other than for the purpose of operating or lighting a ventilating system should be installed in any ventilating duct; similarly, no wiring other than that associated with the lift should be installed in a lift shaft. See **British Standard** 5655 for details.

6.25 Flexible conductors should be as short as practicable and should be provided with mechanical fixing and reinforcement at the point of entry into all apparatus, accessories, fittings or machines to which they are connected.

6.26 Flexible conductors used for arc lamps, floodlights, effects lamps and other electrical apparatus should, where necessary, be protected by heat-resisting material, which in new installations should be cables complying with **British Standard** 6141.

Normal and emergency lighting

6.27 The means of control used for the purpose of providing the **normal lighting** and **emergency lighting** should be situated or arranged so as to minimise the risk of their unauthorised operation.

Normal lighting

6.28 All parts of the **premises** to which **the public** have access and all external exitways should be provided with **normal lighting** capable of providing sufficient illumination of those parts of the **premises** for **the public** to leave the **premises** safely.

6.29 The **normal lighting** should be arranged so that the requirements of paragraph 6.28 continue to be met should a fault occur on other equipment fed from the same source of supply.

6.30 Whenever **the public** are on the **premises**, the **normal lighting** should be kept on, in the absence of adequate daylight, and should provide the level of illumination required by paragraph 6.28, save where **emergency lighting** is in use in accordance with paragraphs 6.31 to 6.42.

Emergency lighting - general

6.31 All parts of the **premises** to which **the public** have access and all external exitways should be provided with **emergency lighting** capable of providing sufficient illumination for **the public** to leave the **premises** safely.

6.32 The **emergency lighting** should be designed, installed, protected, maintained and operated so as not to be affected adversely by other electrical or gas equipment.

6.33 The **emergency lighting** may be supplied from the same source as the **normal lighting** but should also be capable of being powered by an independent supply. The independent supply should be brought into operation immediately and automatically in the event of failure of the normal supply to the **emergency lighting**.

6.34 The independent source of supply to the **emergency lighting** specified in paragraph 6.33 should be of such capacity that it is capable of maintaining the level of illumination required by paragraph 6.31 for at least one hour or such longer period as may be required by the licensing authority.

6.35 It should be the duty of the person in charge of the **premises** at any time to satisfy a representative of the licensing authority that the **emergency lighting** is capable of maintaining the level of illumination required by paragraph 6.31 for at least one hour or such longer period as may be specified in the licence.

6.36 In theatre auditoria complete or substantially complete blackouts which may be required for production reasons may be

permitted but this should not include extinguishing any exit signs which should, at all material times, comply with the requirements of paragraph 6.44.

6.37 Although **emergency lighting** should be provided in all parts of the **premises** to which **the public** are admitted, it is not envisaged that **emergency lighting** should be provided in each individual WC compartment. However, **emergency lighting** should be provided in toilet accommodation which exceeds $8m^2$ in gross area and in toilet accommodation having a gross area smaller than $8m^2$ where no provision for borrowed light has been made.

6.38 All **emergency lighting** installations which conform to **British Standard** 5266: Part 1 should be subjected to the servicing, testing and inspection specified in that standard.

Emergency lighting - by electricity

6.39 All new **emergency lighting** installations should comply with the appropriate provisions, including the certification, testing and servicing provisions of **British Standard** 5266. This requirement also applies when existing **premises** are rewired or installations substantially altered.

Emergency lighting by gas or other means not using electricity

6.40 In **premises** where **emergency lighting** by gas or other means not using electricity is already installed, the levels of **emergency lighting** provided should, where reasonably practicable, be similar to those which would be provided if the system was powered by electricity and installed in accordance with the provisions of **British Standard** 5266.

6.41 Any flame illuminants should be fixed securely and situated well away from woodwork and other combustible materials, and where possible, out of reach of **the public**. They should be protected from draughts.

6.42 The control of individual gas lamps should either be out of reach of **the public** or be fitted with taps that require a key to turn them. There should however be a means provided by which individual lamps or all lamps can be shut-off in an emergency.

Normal and escape lighting for staff and performers

6.43 Any corridor, not included in parts of the **premises** to which paragraphs 6.28 to 6.30 apply, which may be used by staff and performers in case of an emergency, should be provided with **normal lighting** and an **escape lighting** system.

Exit signs

6.44 All exit and directional signs indicating the exits from any part of the **premises** to which **the public** are admitted should (unless they are **self-luminous fire safety signs**) be illuminated by means of the **normal lighting** and the **emergency lighting** at all times when **the public** are on the **premises**.

6.45 Signs or notices of the photo-luminescent type, ie where the active material making up the luminous parts of such signs or notices needs a period of exposure to light before they become visible in darkness, are not acceptable.

Luminous discharge lighting

6.46 Section 10 of the Local Government (Miscellaneous Provisions) Act 1982 and section 98 of the Civic Government (Scotland) Act 1982, as amended by the Fire Safety and Safety of Places of Sport Act 1987, deal with statutory powers which may be exercisable in connection with the provision, position, colour and marking of emergency switches to cut off the electric supply to luminous tube installations. These statutory powers also require notice of intent to install such equipment to be given to fire authorities.

6.47 The supply of electricity to luminous discharge lighting installations on the outside of a **building**, or used within a **building** when **the public** are not admitted, and in which any voltage normally exceeds the voltage prescribed in section 10 of the Local Government (Miscellaneous Provisions) Act 1982 or regulations made under section 98(1) of the Civic Government (Scotland) Act 1982, should be capable of being cut off by one or more firemen's switches or other means of control situated outside the **building** and be accessible to members of the fire brigade but not easily accessible to **the public**.

6.48 In the case of luminous discharge lighting installations where the voltage exceeds that prescribed, the installation should comply with the **IEE Regulations for Electrical Installations** and **British Standard** 559. Any firemen's switches required to control luminous discharge lighting should comply with the above Regulations.

6.49 Transformers, starting and control gear, and any other gear for luminous discharge lighting installations, should be effectively protected by earthed metal enclosures.

6.50 Insulating oil or other substances which have a flash point of less than 120°C, and are fluid at normal working temperature, should not be used in any transformer or other apparatus forming part of an electric discharge tube installation, unless the transformer or other apparatus is protected by special devices which will automatically disconnect it from the supply circuit in the event of any abnormality causing dangerous overheating.

Lighting for special effects

6.51 No lasers, strobes or other form of high intensity lighting should be installed or used without **approval**. No such lighting should be used other than strictly in accordance with any terms and conditions which may be imposed by the licensing authority. (See also paragraphs 11.40 to 11.44.)

Inspection of electrical installations

6.52 All electrical installations should be inspected at least every 5 years or such shorter period as may be specified by the licensing authority, by a competent electrical engineer as described in paragraph 6.2 appointed by, or on behalf of, the **manager** of the **premises**. An inspection certificate concerning the condition of the installations should, after each inspection, be forwarded to the licensing authority by or on behalf of the **manager** of the **premises**. In the case of new installations or extensions to existing installations a completion certificate should be obtained. Specimen completion and inspection certificates are produced by the Institution of Electrical Engineers.

Generators, transformers, switchgear and heat producing equipment

6.53 Electricity generating plant and main supply transformers should be placed in a room:-

(a) other than the auditorium;

(b) not communicating directly with the auditorium or with any other part of the **building** to which **the public** are admitted; and

(c) which is of **fire-resisting construction** throughout (except where there are windows, skylights and openings communicating directly with the open air). The standard of fire-resistance should not be less than 60 minutes or such higher standard as may be required by the licensing authority.

Heating appliances

6.54 All **heating appliances** should be suitably guarded and fixed in position in such a manner so as to prevent unauthorised persons having access to the controls or being able to approach sufficiently close to the appliance to endanger themselves.

6.55 Every **heating appliance** used in the **premises** should be situated at a safe distance from any combustible materials or substances. In deciding the safe distance regard should be given to the type of heater and the ease of ignitability of the nearby materials or substances.

6.56 Care should be taken to ensure that radiant heaters, particularly those fitted with reflectors, are not directed towards combustible materials and that, wherever possible, they are mounted at a high level above head height, usually at least 2.5 m above floor level.

Note: *It is not possible to give definitive advice on the safe distance necessary between heater and combustible material due to the many variable factors involved. Generally any combustible material should not have its temperature raised significantly above the ambient temperature for the area by the effects of direct heating by the appliance.*

6.57 No oil-fired heaters, other than those forming part of the boiler installations, should be used in the **premises**.

6.58 No portable liquefied petroleum gas (LPG) heater should be in the **premises** when members of **the public** are present. Containers of LPG should be protected against unauthorised interference and accidental leakage. LPG cylinders, both full and empty, should be kept in safe positions in the open air away from other flammable materials, or where this is not reasonably practicable because of exceptional circumstances, in an adequately ventilated storeroom constructed in accordance with HSE Guidance Note CS 4 - 'The keeping of LPG in cylinders and similar containers' (HM Stationery Office, ISBN 0 11 883539 4). The storeroom should not be used for any purpose other than the storage of LPG cylinders. LPG cylinders in use should be sited outside the structure and with their valves readily accessible in case of an emergency.

Exclusion and admission of the public

6.59 All entertainments should immediately stop and **the public** instructed to leave the **premises**:-

(a) if there is a suspected gas leak; or

(b) if the **normal lighting** fails and remains inoperative (see paragraphs 6.28 to 6.30) for more than the rated duration of the **emergency lighting** system, less one hour. **The public** should not be re-admitted until the requirements of paragraphs 6.28 to 6.30 can be met; or

(c) if the **normal lighting** fails, and the **emergency lighting** has a rated duration of less than one hour; or

(d) if the capacity of the source of supply for the **emergency lighting** can no longer maintain that lighting for at least one hour; or

(e) if the **emergency lighting** fails to comply with the level of illumination required by paragraphs 6.31 to 6.37; or

(f) in the case of small **premises**, if the **emergency lighting** fails, unless a rapidly available alternative lighting system, meeting the requirements of paragraphs 6.31 to 6.37 and, if necessary, paragraph 6.44 is brought into operation without delay and the **normal lighting** is still operating.

Note: The term 'rated duration' in (b) and (c) above is defined as the period of time that the luminaire can continuously provide the minimum illuminance required in the emergency condition. The time is specified in hours.

Notes/Amendments

Chapter 7: SEATING ARRANGEMENTS

Introduction

7.1 The type of seating arrangements adopted in **places of entertainment** will vary with the use to which the **premises** are put. **Premises** should not normally be used for closely-seated audiences except in accordance with a plan which should have been submitted previously for **approval**. (For disabled persons, see Chapter 2.)

Seating and gangways

7.2 Seating and gangways in the auditorium or hall should be so arranged as to allow free and ready access direct to the exits.

7.3 No portion of any gangway should normally be more than 18m from an exit from the auditorium or hall measured along the line of the gangway. (See diagram 6.)

7.4 In those parts of the auditorium where only one exit is available no portion of any gangway should be more than 7.5m from that exit measured along the line of the gangway.

7.5 The seating assigned to each person should be not less than:–

(a) 760mm deep where backs are provided to the seats or 600mm deep where backs are not provided;

(b) 500mm wide where arms are provided to the seats or 450mm wide where arms are not provided.

7.6 In fixed seats there should be a clear seatway or space of at least 305mm measured between the perpendiculars from the back of one seat to the front of the seat immediately behind it. (See diagram 7.)

7.7 Where seats tip up automatically the clear seatway should be measured between the back of one seat and the maximum projection of the seat behind when that seat is in the 'Up' position. (See diagram 8.)

7.8 The calculated occupant capacity of the **premises** (or part of the **premises**) should be determined:–

(a) in areas where fixed seating is provided:–

(i) if individual seats, by the number of such seats, and

(ii) if bench seats or similar continuous seating, by dividing the total width of such seating by 450mm; and

(b) in other areas (including standing areas occupied together with fixed seating) by dividing the floor area in metres² by the relevant occupant load factor given in Table C of paragraph 5.20.

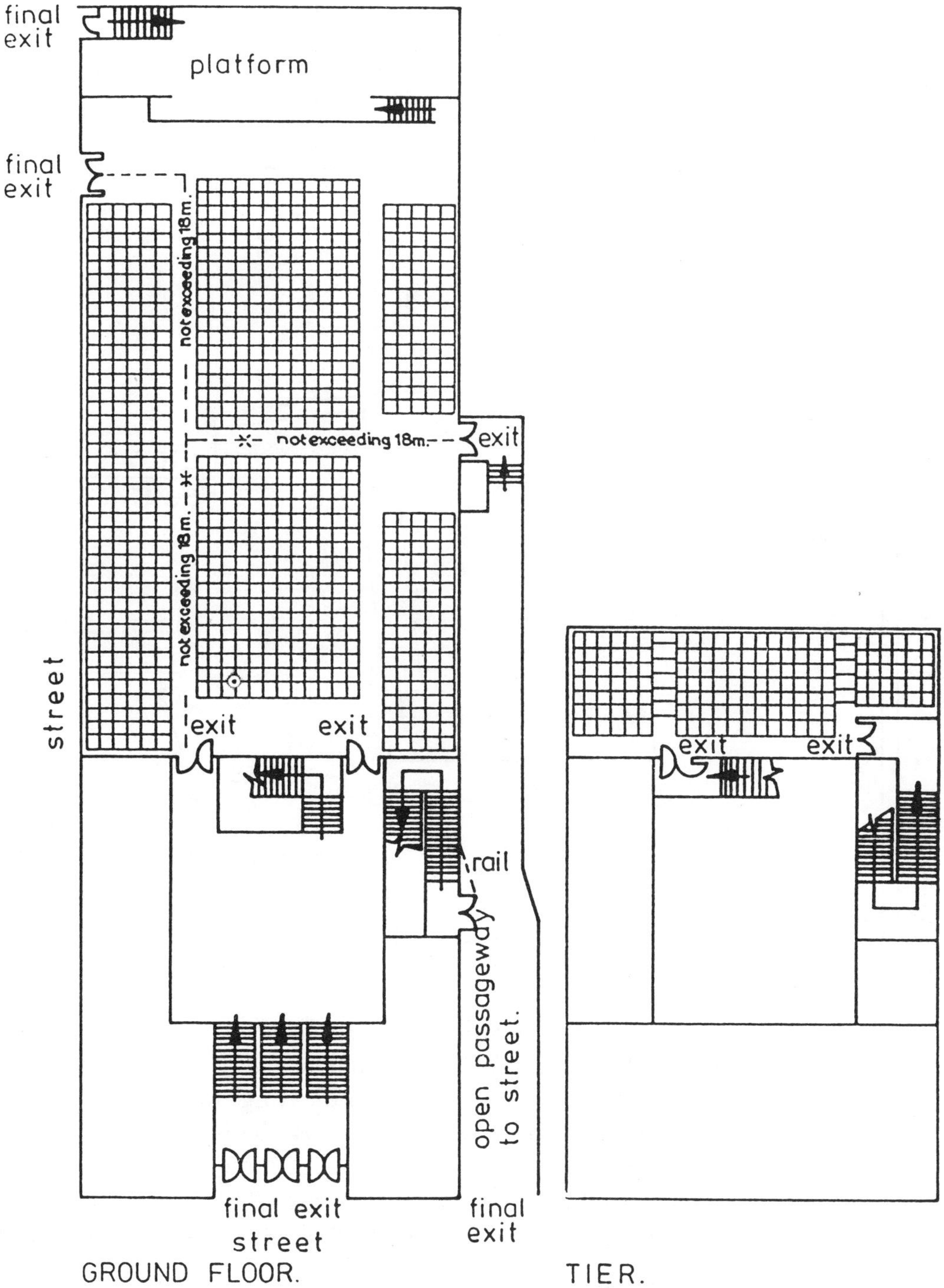

Diagram 6 Example of assembly hall on sloping ground floor for 750-1000 persons and 150 in stepped tier showing distribution and separation of exits, arrangements of gangways and blocks of seating.

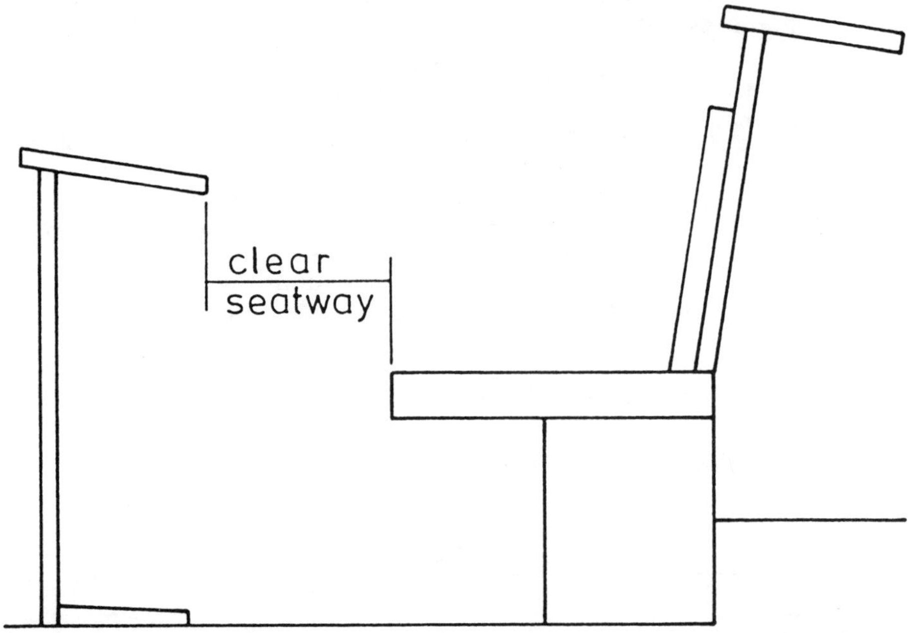

Diagram 7 Example of fixed seats.

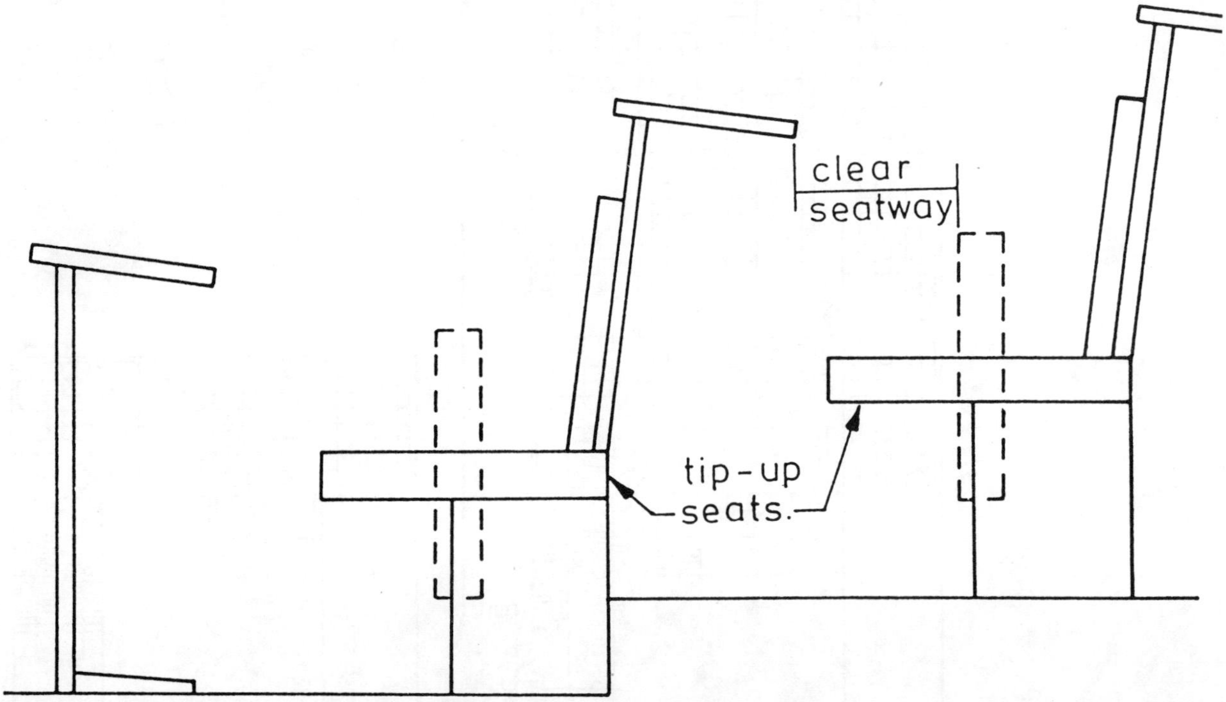

Diagram 8 Example of tip-up seats.

7.9 It is the practice of many licensing authorities to require that the standing area should be marked with a white line or, alternatively, with a conspicuous notice indicating the number of persons **permitted** to stand there. Either practice is recommended.

7.10 Gangways should be of adequate width for the number of seats served but should in no circumstances be less than 1.05m wide.

7.11 There should be no projection into the gangway which would diminish its clear width, and the ends of all rows and seats should be so aligned as to maintain a uniform width of gangway throughout its length.

7.12 A continuous handrail should be fixed on each side of all stairs, steps, landings and ramps at a height of not less than 840mm or more than 1m, measured vertically from the pitch line of the nosings of steps and from the floor of landings, except that a second handrail need not be provided to such stairs and ramps or flights of steps which are not more than 1.05m wide. The handrail should not normally project more than 100mm and the ends of the handrail should be turned for safety.

Conventional seating

7.13 Subject to the provisions outlined in paragraph 7.3, the number of seats in a row should not exceed:–

(a) 7 seats where there is a gangway at one end only except that up to a maximum of 11 seats may be **permitted** if the 305mm seatway referred to in paragraph 7.6 is increased by 25mm for each additional seat over 7; and

(b) 14 seats where there is a gangway at each end except that up to 18 seats may be **permitted** if the 305mm seatway referred to in paragraph 7.6 is increased to not less than 400mm, and more than 18 seats may be **permitted** if the seatway is not less than 500mm.

Continental seating

7.14 Continental seating (see diagram 9) refers to rows in which there are more than 22 seats. This arrangement of seating may be **permitted** subject to the following additional requirements to those detailed in 7.13 (b):–

(a) no seat should be more than 15m from an exit measured along the line of travel; and

(b) gangways or exits should be provided at each end of a row of seats. Where gangways are provided, the position of the exits and seating should be arranged so that the streams of persons leaving the seatway move in the gangways in a direction away from the **stage** or **platform.**

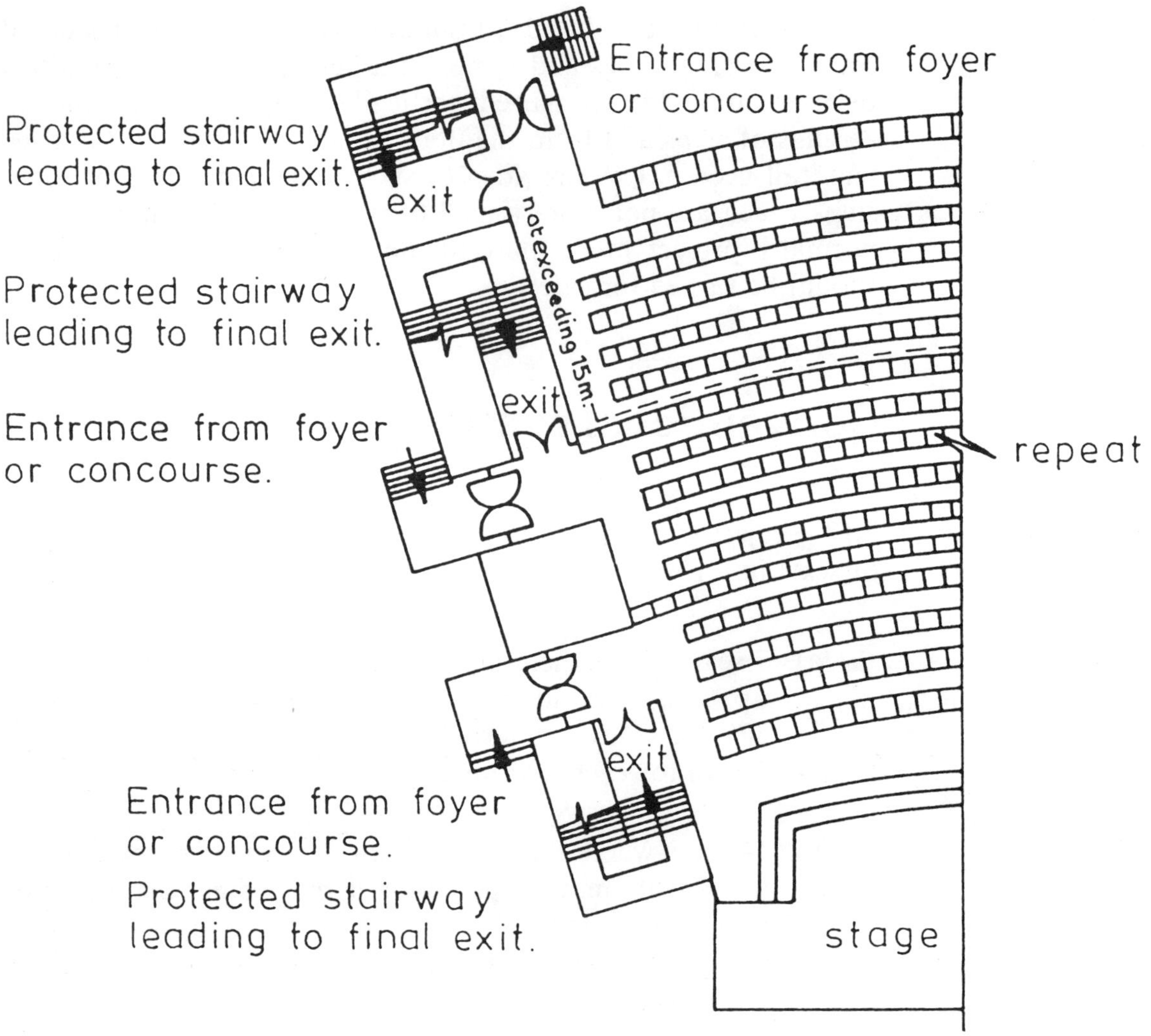

Diagram 9 Example of Continental seating and exits.

Diners' seating and lounges

7.15 Where the audience is seated at chairs around tables, the seating need not be fixed, but both the tables and chairs should be arranged so that clearly identifiable gangways are provided and conform with the relevant occupant load factors given in Table C of paragraph 5.20.

Peripheral seating

7.16 This seating which includes seating alongside an arena such as a skating rink or a dance floor need not be fixed provided that:–

(a) it comprises individual seats in the form of not more than 3 rows; or

(b) it is random seating within an area not more than 3m in depth; and

in either case the gangways to the exits are kept clear of obstruction by the provision of fixed barriers designed to prevent the encroachment of the seating into the gangways.

Seating arrangements

7.17 Where permanent provision is made for a closely seated audience and in *all* cases of stepped tiers, *all* seating should be fixed firmly to the floor (the only exception to this being chairs in boxes or other **approved** enclosures) (but see Chapter 2 about seating for the disabled).

7.18 In areas within **premises** where there is normally no fixed seating, any seating specially provided for an indoor concert should be so arranged and fixed in position, that it cannot be moved easily by an audience in a state of excitement.

7.19 In circumstances where more than 250 temporary seats are to be used in the auditorium the following arrangements should apply:–

(a) chairs or other single seats should be secured together in lengths of not fewer than 4 seats and not more than 12 so that the seats cannot be separated from each other merely by pushing one or more seats in the row (the only exception being chairs in boxes or other **approved** enclosures); and

(b) provision should be made for the rows of seating flanking the gangways to be fixed to the floor effectively preventing the individual seats or rows of seats from being dislodged into the gangways or from being toppled over, except that:-

(i) only end seats of such rows need to be fixed to the floor if all seats in each row are secured together; or

(ii) only the end seats of each length of seating referred to in paragraph 7.19(a) above which form such rows need be fixed to the floor.

*Note: In **premises** which are intended to be used only occasionally for closely seated audiences where the fixing of seating to the floor is impractical or undesirable (eg on polished dance floors) the use of floor bars instead of floor screws may be **permitted**. Such floor bars should be not more than 25mm in height, have a cambered top surface so as to avoid the risk of tripping persons using the seatways, and should extend from the row to be fixed to at least two adjacent rows, but should not extend across any gangways. (See diagrams 10, 11 and 12 for examples of fixing arrangements.)*

Seating at indoor sports events

7.20 Seating at indoor sports events should be in accordance with paragraphs 7.13 and 7.14.

Seating at pop concerts and like occasions

7.21 It is recommended that pop concerts and other like occasions which are likely to attract mainly younger audiences should be held only in **premises** (such as concert halls and theatres) where the audience can be separated into manageable units by means of adequate fixed seating and, so far as is practicable, by guardrails and protective barriers (see paragraphs 5.82 to 5.85).

7.22 In **premises** with permanently fixed seating, which are used for pop concerts for mainly younger audiences, consideration should be given to the permanent provision of suitably designed hinged barriers across gangways to assist in the control of the audience. Any such barrier should be provided only after consultation with the licensing authority and, where installed, should be fixed in such positions that they effectively control any movement of the audience towards the **stage** but at the same time are not a hazard to the audience in the event of an evacuation of the **premises**. They should be manned by attendants at all times.

7.23 The space in front of the front row of the seats at stalls level should be treated as a gangway, and in venues which are regularly used for pop concerts, it is recommended that this gangway space be reduced at the front to control the crowd and prevent ready access to the **stage area**.

7.24 Where a pop concert takes place at **premises** which are comprised of an outdoor sports stadia it should be appreciated that although part of the audience may be seated, a large part of the audience will be on the actual sporting arena and that this mobile audience may provide additional problems in respect of **means of escape**, adequacy of exits etc. Full consideration therefore needs to be given to the occupant load factor (see paragraph 5.20) in relation to the safe evacuation of all present and there should be close co-operation between all enforcing authorities. (See also paragraphs 1.4, 1.14, 1.16, 1.17, 5.20, 5.82 to 5.85, 7.18, and 8.2.)

Seating at outdoor sports stadia

7.25 Guidance on seating arrangements at outdoor sports stadia is contained in the Guide to Safety at Sports Grounds (HM Stationery Office, ISBN 0 11 340840 4).

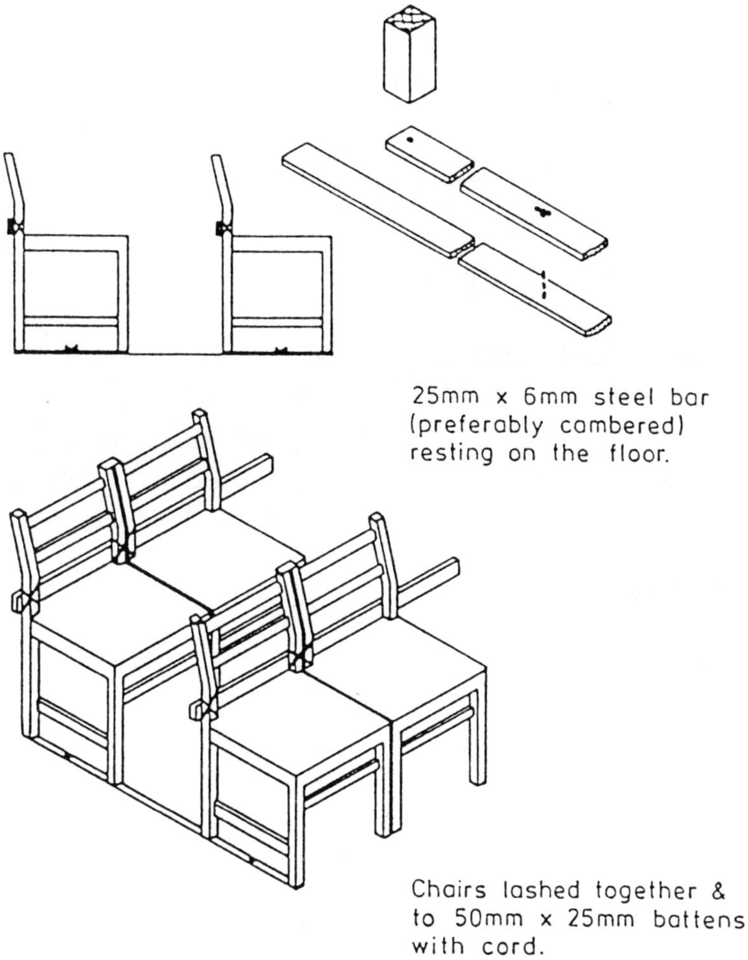

Diagram 10 Example of securing wooden chairs.

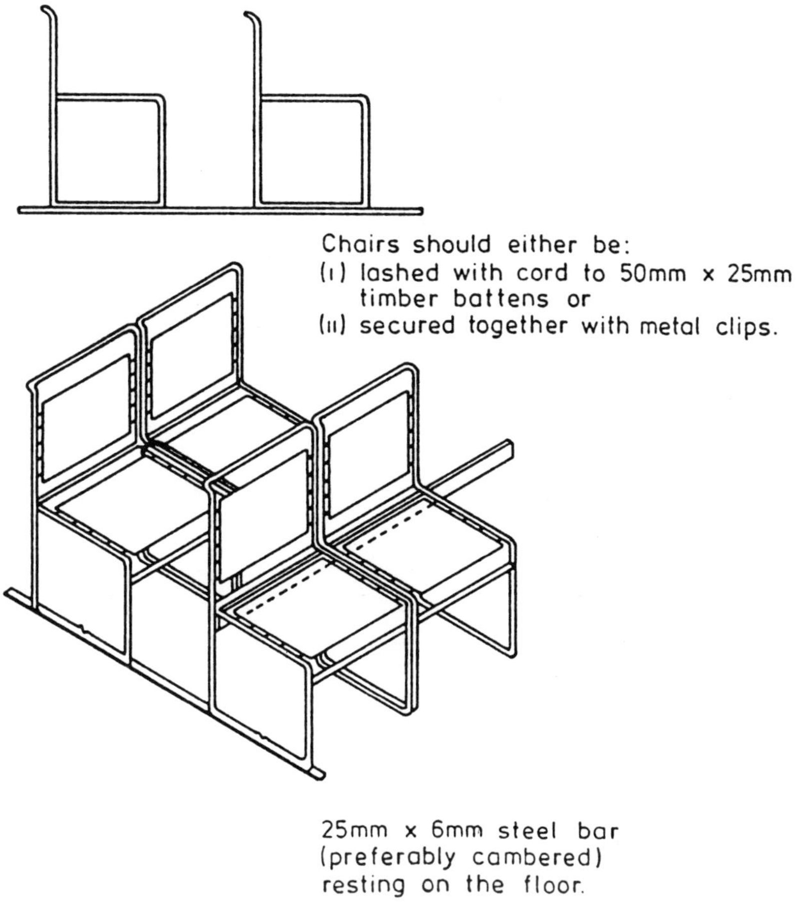

Diagram 11 Example of securing metal frame chairs to floor bars.

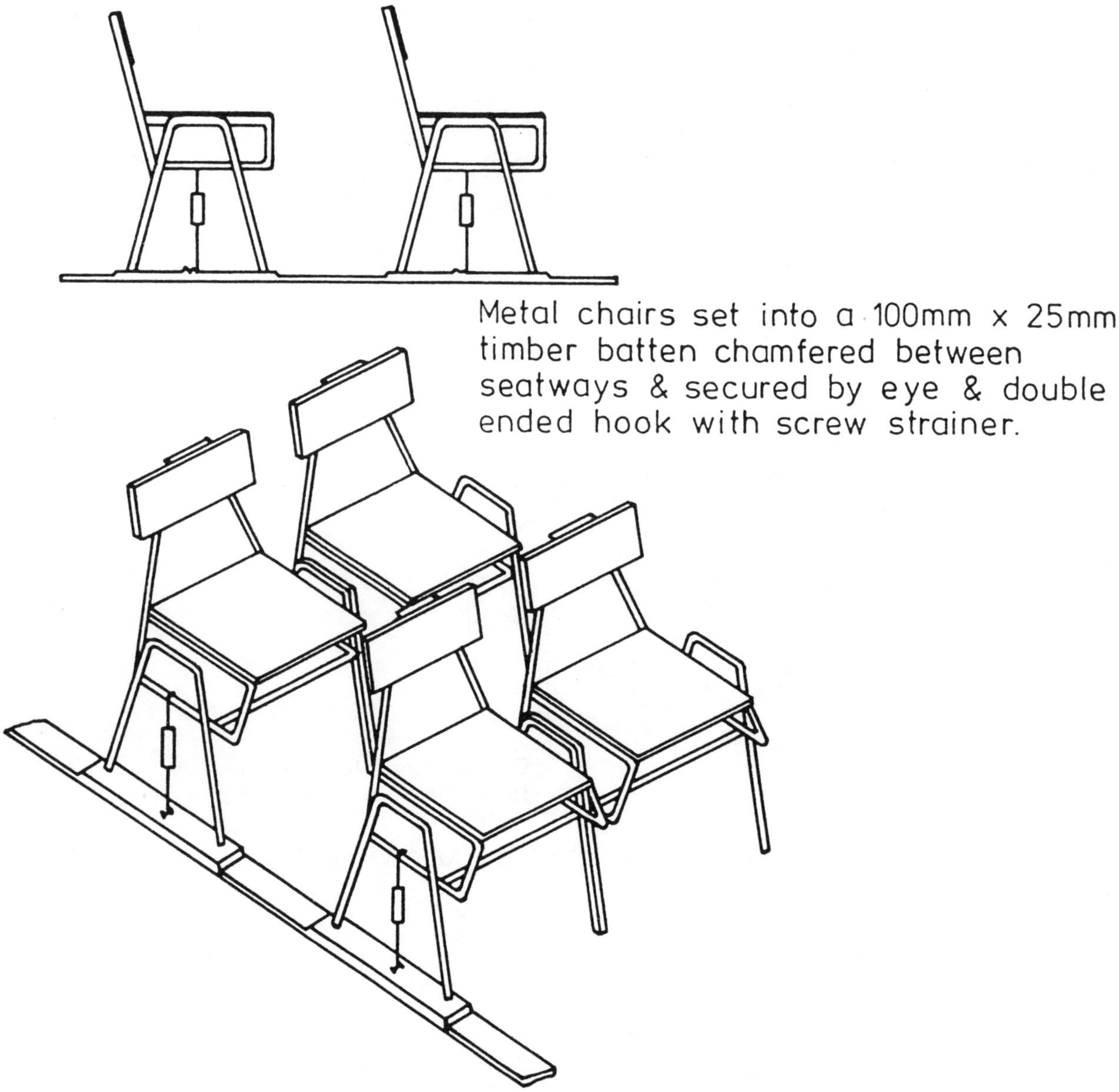

Diagram 12 Example of securing interlocking metal chairs on to a timber floor bar.

Notes/Amendments

Chapter 8:

REQUIREMENTS IN PREMISES HAVING A STAGE

8.1 This chapter deals with **separated stages** and **open stages**. It also deals with **premises** in which a permanent **stage** is not provided.

Separation between audience and stage

8.2 Where the audience may try to get on to the **stage** (eg at a pop concert) the licensing authority may wish to consider whether:–

(a) **stage** boxes close to the **stage** should be prohibited from use; and

(b) a barrier rail or other means of separating the audience from the **stage** should be provided. (See also paragraph 8.27.)

Separated stage

8.3 A **separated stage** is one that is separated from the auditorium and the rest of the **premises** and will include the provision of a safety curtain at the proscenium opening. There should be a means for releasing smoke and hot gases in the event of fire in the **stage area**. (See diagram 13.)

Safety curtain

8.4 The proscenium opening should be provided with a safety curtain which should be of robust construction, consisting entirely of **non-combustible materials** and should be able to withstand damage by **scenery, properties** or falling debris. It should:–

(a) be of sufficient strength and stiffness to resist the pressure of air likely to be caused by fire in the **stage area** without distorting to an extent which could cause its withdrawal from its retaining guides;

(b) provide an adequate seal against the passage of smoke between the movable curtain and the elements of structure and safety curtain guides;

(c) be able to withstand the effect of fire for a sufficient period of time to allow the complete evacuation of the **building**; and

(d) be capable of closing the opening within 30 seconds.

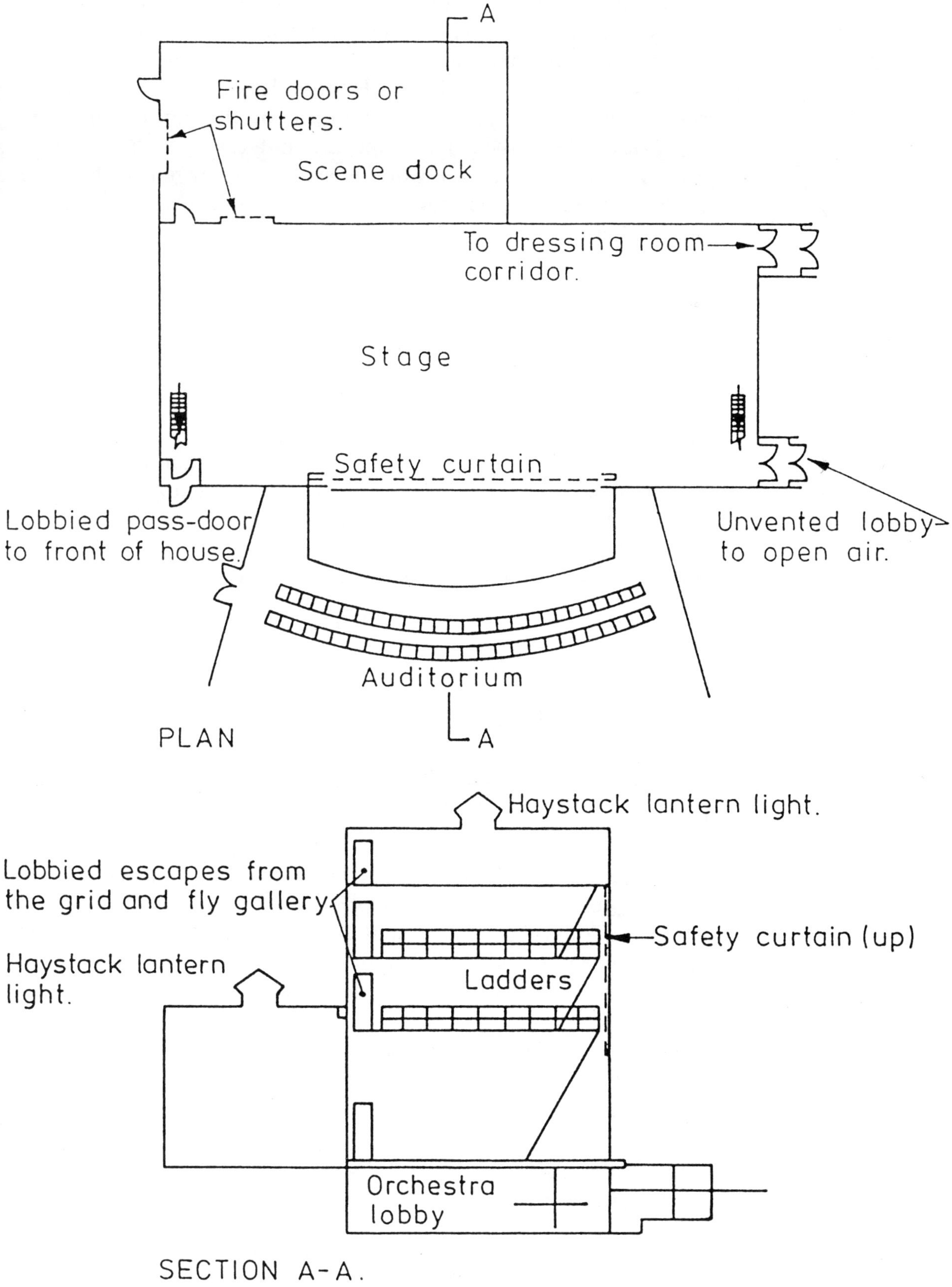

Diagram 13 Example of a separated stage showing exits, smoke-outlets and separations from flame-retarded scenery.

8.5 The words 'SAFETY CURTAIN' should be painted conspicuously on the curtain so as to be clearly visible to the audience from all parts of the auditorium.

8.6 The curtain and the curtain guides should be protected by a manually released drencher system arranged to spray the whole of the **stage** face of the curtain with water and to keep such curtain and guides cool in the event of fire. Hand release gear, to cause the descent of the curtain and the operation of the curtain drencher system, should be provided in duplicate and clearly indicated. One such release should be on the working side of the **stage** and the other in a position outside the **stage** readily accessible to firefighters or authorised staff.

Stage ventilation

8.7 Means should be provided for ventilating smoke and hot gases resulting from fire on the **stage**. This may be achieved by the provision of a haystack lantern light or fire ventilator sited in the highest point in the roof over the **stage** and as near to the centre of the **stage** as is reasonably practicable. To be effective, provision may need to be made for an additional fresh air inlet.

Haystack lantern light

8.8 The base of any lantern light should have an area not less than 10 per cent of that of the **stage** .

8.9 The sashes on each side of any lantern light should be bottom hung so as to open outwards and be glazed with thin sheet glass. However, as this glass could shatter inadvertently, consideration should be given to the use of a thin alloy or plastic sheet which would rupture or melt when subject to fire conditions.

8.10 The aerodynamic free area of the sashes on each side of a lantern light, when open, should be not less than 10 per cent of the area of the **stage**.

8.11 Provision should be made for all sashes in any lantern light to be opened simultaneously:–

(a) on the operation of a fusible link designed to operate at a temperature not exceeding 74°C and so sited below the base of the lantern light as to be clear of the water spray from any sprinkler or drencher system provided; and

(b) on the operation of manual releases which should be sited adjacent to the safety curtain and drencher releases.

8.12 Where the lantern light is provided with a glass roof, the roof should be of a wired glass construction.

8.13 Safety chains may be fitted to the sashes provided that they allow the sashes to open to the full extent required in paragraph 8.10.

Fire ventilator

8.14 Where a fire ventilator is provided it should have a rated aerodynamic free area not less than 10 per cent of that of the **stage**. Where more than one ventilator is fitted they should have a combined area not less than that required for a single ventilator. Provision should be made for each ventilator to open simultaneously:–

(a) on the operation of a fusible link sited in the inlet, designed to operate at a temperature not exceeding 74°C and so sited as to be clear of the water spray from any sprinkler or drencher system provided; and

(b) on the operation of manual releases which should be sited adjacent to the safety curtain and drencher releases.

Maintenance etc

8.15 Facilities should be provided to enable the openable sashes to lantern lights and the mechanisms of fire ventilators to be regularly maintained. In addition, means should be provided whereby the sashes and ventilators can be closed following a test without recourse to the roof.

Note: *The aerodynamic free area of a ventilator is the product of the measured free area in the throat multiplied by the discharge coefficient. For louvered ventilators the discharge coefficient is normally in the order of 0.65, which takes into account any obstructions in the ventilator throat such as louvres and controls etc.*

Scenery store

8.16 A **scenery store** should be separated from the **stage** and auditorium by **fire-resisting construction**.

8.17 Any opening from a **scenery store** to the **stage area** should be fitted with a single **fire door** or shutter which should be self-closing on the operation of a fusible link and any other fire detection system installed. The door or shutter should be kept shut when **the public**

are in the **building** and at all other times, except when **scenery** is being moved through it.

8.18 Means should be provided for ventilating smoke and gases from a fire in the **scenery store**. The aerodynamic free area of ventilation provided should be not less than 10 per cent of the floor area of the **scenery store**. The manual operating device should be sited in a readily accessible position outside the **scenery store**. To be effective, provision may need to be made for an additional fresh air inlet.

8.19 A reduction in the standards set out in paragraphs 8.16 to 8.18 may be **accepted** provided that:–

(a) the auditorium for **the public** is on not more than two floors and is for not more than 500 persons;

(b) there are reasonable precautions to counteract the dangers posed by a fire on or in the areas surrounding the **stage area**;

(c) provision is made for flame-retarded curtains to separate the **stage** from the auditorium; and

(d) the **means of escape** arrangements are satisfactory.

Open stage

8.20 An **open stage** is one that is not separated from the auditorium but is separated from the rest of the **premises**, in consequence of which additional requirements relating to **scenery**, special effects, etc may be necessary.

8.21 Scenery or **properties** should not be kept or used on an **open stage** other than in an **approved scenery store** or **property** store.

8.22 The use or storage, on an **open stage**, of the following may be **approved**:–

(a) **non-combustible material;**

(b) **inherently flame-retarded fabric**;

(c) timber, hardboard or plywood rendered flame-retarded by an **acceptable** process of impregnation (see paragraph 8.50);

(d) timber framing not less than 22mm thickness;

(e) **durably flame-retarded fabric**;

(f) fabric rendered and maintained flame-retarded to the satisfaction of the licensing authority;

(g) blockboard, plywood or chipboard not less than 18mm minimum thickness; and

(h) subject to special consideration, **acceptable** plastics material. (See paragraphs 8.51 and 8.52.)

8.23 The licensing authority may make special conditions governing the use of **properties**.

General requirements for both permanent and temporary stages

8.24 The **stage** should be constructed and arranged so as to minimise the risk of fire.

8.25 For a permanent **stage**, the floor should be constructed of tongued and grooved boarding, not less than 29mm in finished thickness, and should be sufficiently strong to support any type of performance which is likely to be given. The use of suitable plywood of sufficient thickness may be **accepted** in lieu of tongued and grooved boarding. A temporary **stage** may be constructed to a lower standard to the satisfaction of the licensing authority.

8.26 The supports or supporting structure to a **stage** floor should be constructed of such materials and in such manner as the licensing authority may consider necessary.

Note: *Where any space under the **stage** is enclosed to the licensing authority's satisfaction and is not to be used for any purpose other than the accommodation of lifting or other mechanism ancillary to the **stage** or as a means of access by staff or performers, the structure supporting the **stage** may be of unprotected metal or stout timber.*

Pop concerts and like occasions

8.27 Where **premises** are used on a regular basis for pop concerts and like occasions, the licensing authority may decide that the **stage** should be of sufficient height to deter individuals from trying to climb on to it. They may also consider that a safety barrier should be provided around the **stage** to prevent access to the **stage** itself. Any barrier should be constructed to the satisfaction of the licensing authority. (See also paragraph 8.2.)

Grid and galleries

8.28 The grid and galleries including lighting galleries and perches should normally be of **non-combustible** construction except that superimposed walkways to galleries may be of other **approved**

material. In older **premises** the structural members of the grid and galleries may be **acceptable** in timber if of substantial size.

Suspension of scenery

8.29 The means of suspension for **scenery** should be **acceptable** as regards loading, position control and guarding of the mechanism and cables. The construction should, as far as practicable, be **non-combustible**.

8.30 Flown **scenery** should preferably be suspended from wire ropes, attached either to counterweights or self-sustaining winches, so as to retain full control of the position of the **scenery** while in use. Any bars used for the attachment of **scenery** should normally be of metal and suspended by normally not less than 3 wire ropes. Adequate provision should be made for adjustment to offset stretch of these ropes.

8.31 Suspension ropes should be of flexible steel, should have a safety factor of not less than 6, and should be carried on pulley sheaves of diameter not less than 30 times the rope diameter and mounted on free-running bearings. Handling ropes should be of hemp, nylon or similar man-made fibre.

8.32 Counterweights of **scenery** should be guarded, and where possible, should be carried to the walls and cased in.

Exits from stages and associated areas, flies and grids

8.33 Exits to afford ready **means of escape** from all parts of the **stage area** should be arranged as follows:–

(a) at least 2 exits should be provided from the **stage.** Where required by the licensing authority, one of these exits should lead direct to a street, way or open space, through an unventilated lobby of solid construction having a standard of fire resistance of not less than 30 minutes and having **fire doors** at each end. All exit doors from the **stage** should be free from fastenings other than panic bolts;

(b) if the **stage** basement is to be used in connection with a performance it should have at least 2 exits, one of which should be independent of the **stage** . The doors should be free from fastenings other than panic bolts;

(c) unless otherwise **permitted** all dressing-rooms should be connected with **means of escape** leading direct to a street, way, or open space, and one of the **means of escape** should be independent of any **means of escape** by way of the **stage**;

(d) if considered necessary by the licensing authority, alternative **means of escape** should be provided from **scenery stores**, **property** stores, wardrobes and painting-rooms, and all escape doors should be free from fastenings other than panic bolts;

(e) access from the **stage** to the flies (including the lighting flies and perches, other than lighting perches not more than 1.2m above or below a platform to which suitable access is provided) and to the grid should be provided by means of step-ladders with handrails. Where possible the use of cat ladders should be avoided; and

(f) the working fly gallery and grid should be provided with alternative **means of escape** either direct to the open air, or to another part of the **premises** (other than the auditorium) through a **protected lobby** via a self-closing 30 minute **fire door.**

8.34 The exits from the **stage** should be clearly indicated by exit notices which should comply with the requirements of **British Standard** 5499: Part 1 and either be **self-luminous fire safety signs** complying with the requirements of **British Standard** 5499: Part 2 or be illuminated by 2 independent systems of lighting. Internally illuminated signs should comply with the requirements of **British Standard** 2560.

8.35 The routes of escape from the dressing-room block, and from **scenery stores**, **property** stores, wardrobes and painting-rooms should be suitably indicated by such 'EXIT' and 'WAY OUT' notices and directional signs as are considered necessary by the licensing authority.

8.36 Any **scenery store**, **property** store, wardrobe or painting-room should be separated from the **stage** and from the remainder of the **premises** in an **approved** manner.

*Note: Except for any openings which are **permitted** the separation should be of construction having a standard of fire-resistance of not less than 2 hours, provided that where 2 or more enclosures used for similar purposes are adjacent, the standard of fire-resistance of the separation between such enclosures need not exceed 30 minutes.*

Under-stage areas

8.37 Any area under the **stage** which may be used in conjunction with a **stage** presentation should be regarded as part of the **stage** and should have at least 2 exits, one of which should be independent of the **stage.**

8.38 Any area under a **stage** used for other purposes should be separated from the remainder of the **building** by 60 minutes **fire-resisting construction.**

*Note: No storage of combustible materials should be allowed under a temporary **stage**.*

Guidance on scenery or properties used for stage presentations

8.39 The **licensee** should give the licensing authority at least 7 days' notice in writing of the date(s) on which it is intended that the production will take place, and as much notice as possible as to when **scenery** and **properties** will be available for inspection and any necessary testing. The licensing authority should also be notified before **scenery** and/or **properties** are introduced into a current production.

8.40 All **scenery** including cloths, draperies, gauze-cloths, floral decorations, hangings, curtains and all fabric decorations on the **stage** should be maintained flame retarded. (For further advice on artificial and dried foliage see paragraphs 4.20 to 4.23.).

8.41 Scenery or **properties** should not be stored or kept in the **stage** basement except when required for use in a current production. **Scenery** or **properties** should not be kept or used in any other part of the **premises** other than the **stage** or in an **approved property** or **scenery store**.

8.42 Scenery or **properties**, other than those required for use in the current production, should not be kept on the **stage** without **approval. Approval** may be given where **scenery** or **properties** are required for use in rehearsals of impending productions and also to those **premises** where **acceptable** permanent provision has been made to accommodate additional **scenery** or **properties.**

8.43 The **platform** or **stage** should not be congested with **scenery** or **properties** and the exits leading from the **stage** should be maintained free from obstruction.

Testing of scenery

8.44 Fabrics etc, which need to be rendered flame-retarded by means of a flame-retardant treatment should preferably be so rendered during the process of manufacture and by a process giving durable flame-resistance.

8.45 The licensing authority will not normally require any particular method to be adopted to render materials flame-retarded (except in so far as it distinguishes between durable and non-durable processes). (See technical definitions of **durably** and **non-durably flame-retarded fabric**).

8.46 Materials treated by a non-durable process which have been washed or cleaned should be re-treated to render them flame-retarded before re-use.

8.47 Where a non-durable process is **permitted** a range of proprietary products are available for the treatment of materials.

8.48 Electrical wires or apparatus should be removed before any material is submitted to a process of flame-retardant treatment.

8.49 Durably flame-retarded fabric may be **accepted** for **scenery** etc on an **open stage** but the licensing authority may wish a valid certificate to be produced confirming that it is flame retardant and detailing the treatment. The licensing authority may also wish to satisfy themselves at the time of inspection that the fabric does not readily burn.

8.50 An '**acceptable** process of impregnation' for timber scenery (see paragraph 8.22(c)) means a vacuum/pressure process (whereby the salt-retention in the material can be controlled) undertaken by a processor whose products are **acceptable**. Such products will bear a distinguishing mark to indicate they have been so impregnated (the standard for such material being Class 1 when tested in accordance with the provisions of **British Standard** 476: Part 7).

8.51 'Acceptable plastics material' means thermoplastic material which, when tested:–

(a) in accordance with Method 508A of **British Standard** 2782: Part 5, is such that three consecutive specimens do not burn to the first mark and whose flame or after glow, after removal of the burner, does not endure for more than 5 seconds; and

(b) in accordance with either Method 508C of **British Standard** 2782: Part 5, or with **British Standard** 2782: Part 1: Method 140D, is such that the distance the flame travels is not more than 75mm; or

(c) in accordance with either Method 508D of **British Standard** 2782: Part 5, or with the current edition of **British Standard** 2782: Part 1: Method 140E, is such that for each of the 3 specimens, or 5 out of 6 specimens:–

(i) the length of time the specimen glows or flames does not exceed 5 seconds;

(ii) any material which may drop from the specimen does not continue to burn after reaching the base of the test apparatus;

(iii) the percentage of the area of the underside of the specimen that is charred or scorched does not exceed 20 per cent; and

(iv) the length of that part of the edge on the underside of the specimen that is charred or scorched does not exceed 50mm.

*Note: If a **licensee** wishes to use a plastic material which falls outside of the above criteria it will be their responsibility to satisfy the licensing authority of its suitability by providing a test certificate from a body of recognised standing.*

8.52 When a material achieves the standards set out in Chapter 8 but falls below Class 1 when tested in accordance with **British Standard** 476: Part 7, the licensing authority would normally consider imposing a limitation as to the position and/or quantity of material to be used.

Scenery for entertainment in premises not fully equipped as theatres

8.53 Where **scenery** is used in **premises** not fully equipped as a theatre it will be for the licensing authority to decide which conditions or limitations should be applied.

8.54 The amounts and type of **scenery** should be restricted according to the arrangement of the **premises**. Whilst consideration should be given to special circumstances or arrangements at individual **premises**, the general standards set out in paragraph 8.56 should be achieved.

8.55 Flame-retarded scenery additional to the quantities specified may be **permitted** in **premises** where there is a properly arranged **scenery store** or where the licensing authority is satisfied that all **scenery**, including that not actually in use, can be placed so as not to obstruct access to the **stage** exit, fire appliances, safety equipment releases or switchgear.

8.56 Where an excess of **scenery** or the use of special effects are **permitted** compliance with the following special conditions may be required:–

(a) in **premises** having a brick or other suitable proscenium, **durably flame-retarded** tab curtains, exits satisfactory as to number and disposition from both **stage** and auditorium, and satisfactory **means of escape** from any grid or fly galleries, the flame-retarded scenery kept or used on the **stage** should not normally exceed at any one time 3 backcloths (2 of which should be kept rolled while the third is in use), 8 wings and 3 borders, or their equivalent as 2 box sets;

(b) in **premises** having a proscenium and/or other arrangements less satisfactory than those described above, the flame-retarded scenery kept or used on the **stage** should not normally exceed at any one time 2 backcloths (one of which should be kept rolled while the other is in use), 4 wings and 3 borders, or their equivalent as a box set. Any additional **scenery** required in the production should be constructed from such materials which are **accepted** for use on **open stages**.

Note: For small ***premises*** *see Chapter 12.*

Notes/Amendments

Chapter 9:

MEANS FOR GIVING WARNING IN CASE OF FIRE

General

9.1 In any **building** (or part of a **building**) used as a **place of entertainment** there should be a means for giving a warning in case of fire to persons in the **building**.

Fire alarm standards

9.2 A fire alarm system should comply with the recommendations for Type M of **British Standard** 5839: Part 1 and where an alarm system is installed to comply with the **British Standard** a completion certificate should be provided. There may, however, be situations where a simpler means of raising the alarm could be regarded as adequate.

Manually operated alarms

9.3 Manually operated sounders may be **acceptable** in small **buildings** if each sounder will give a warning which is audible throughout the **building** during a performance. Gongs, triangles, klaxons, bells and whistles may be **acceptable** as sounders provided that they cannot be confused with any entertainment in progress.

9.4 Manual fire alarms should be mechanically reliable, and the position chosen for the operation of any sounder should be as safe as possible from the effects of any fire risk, ie away from storage spaces, kitchens, switch cupboards etc.

Call points

9.5 Even in small **premises** it will normally be found to be more satisfactory to install an electrical system, manually operated from 'break-glass' call points situated on escape routes. Call points should be provided on a scale which allows any person to reach a call point without travelling more than 30m. Manual call points should comply with **British Standard** 5839: Part 2.

Staged fire alarms

General

9.6 Research of large-scale fires in public **buildings** in which large numbers of people have been killed has revealed that a major problem is often a delay by staff in warning **the public** of a potential fire risk, combined with a rapid fire and smoke spread. In recommending a staged alarm system it is extremely important to stress that *people need clear and accurate information as early as possible in order to ensure that all are able to reach a* ***place of safety***. A staged alarm system while intended not to alert **the public** unnecessarily, should *not* cause an unnecessary delay in warning **the public** to leave and provision should be made for sounding a general alarm for evacuation by sounders or a signal broadcast over the public address system. The evacuation alarm should be under the control of the **manager** who should decide when the second stage should be put into effect. The staff should be fully aware of the evacuation procedure and the need to direct **the public** to the safest exits. The evacuation plan and staff training should also emphasise the importance of an efficient network of communication between staff members prior to and during an evacuation. (See also Chapter 1.)

9.7 Any warning system involving a staged alarm or phased evacuation should comply with **British Standard** 5839: Part 1 and should be **approved** before installation.

Staff alarms

9.8 In certain **places of entertainment** an initial general alarm may be undesirable because of the number of public present and the need for fully trained staff to effect pre-planned procedures for safe evacuation. It may therefore be preferable to use restricted alarms such as sounders or other types of communication (including personal paging systems) which can be supplemented by visual systems for staff recognition.

Warning signals

9.9 The signals incorporated into the electrical alarm system should be distinctive and the warning devices should be so sited as to ensure that a common warning signal is perceptible throughout the **building** or throughout all parts in which there is a requirement to provide a warning system. In **premises** where the noise level may be excessive, or in any other situation where a normal type of sounder may be ineffective, visual signals should be used to supplement the audible alarms. They should not, however, be used on their own.

Visual and audible alarms

Audible alarms by intercommunication or public address equipment

9.10 Where intercommunication or public address equipment is used in lieu of conventional sounders to transmit a general alarm, the signal should take priority and override other facilities of the equipment. The alarm signal, which may be followed by a voice transmission of essential information for safe evacuation, should be distinct from other signals which may be in general use on the system. Clause 9.12 of **British Standard** 5839: Part 1 gives advice on general requirements.

9.11 Where the normal use of the **premises** is other than for entertainment and a permanent audible fire warning system is installed, it may be necessary for the initial audible alarm to be silenced when parts of the **premises** are used for such entertainments as **stage** plays, etc.

9.12 Audible warning should be provided in the parts of the **premises** to which **the public** are not admitted.

9.13 In order that the audible warning may be silenced in those parts of the **premises** used by **the public** whilst the entertainment is in progress, a switch should be provided in a suitable position which, when operated, renders the sounders inoperative and automatically switches warning signal lamps into the alarm circuit. Control and indicating equipment should show that the audible alarm in that zone is inoperative. The switch should be secured against unauthorised interference and be provided with a notice to denote its purpose together with a coloured indicator lamp, incorporated in the two way switch circuit, to give a visual indication that the alarm sounders have been silenced.

Visual alarms

9.14 Visual alarms for warning staff are recommended for the parts of the **premises** to which **the public** have access. Buzzers or other low noise emission sounders which are meaningful to staff may be used to draw attention to the visual alarm.

9.15 In some **premises**, the **accepted** procedure for notifying staff of an emergency will not rely on a specific and specially installed visual alarm system, but on, for example, the operation of the house lights or, in the case of a theatre, the lowering of the safety curtain, or a code message broadcast over the public address system.

9.16 Arrangements for staff alarm signal lights should be made in **premises** having a closely seated audience such as a theatre, conference or concert hall as follows:–

(a) signal lights should be installed where they will be clearly seen, to warn the staff that a state of emergency has arisen;

(b) the siting of the lights will depend upon the layout of the **premises** but consideration should be given to siting the lights in the following positions:–

  (i) at the back of each tier in the auditorium;

  (ii) in the **manager's** office;

  (iii) in the main entrance hall;

  (iv) near the point where the house lights are controlled;

  (v) on the **stage,** other than an **open stage;**

  (vi) in the **stage** door keeper's office (if provided);

  (vii) near **stage areas**, eg dressing rooms, meeting rooms, staff rooms, workshops etc;

  (viii) in any projection or lighting control box; and

  (ix) adjacent to the doors outside the auditorium where staff may be located.

The lamps used for this purpose should be readily distinguishable by the staff from the decorative lighting of the **premises**. They should not, however, attract the attention of the audience.

9.17 Operating points for all staff alarm signal lights should be placed as follows:–

(a) in or near the **manager's** office;

(b) at each level in the auditorium;

(c) on or near the **stage**;

(d) in or adjacent to the main entrance hall; and

(e) in any projection or lighting control box.

9.18 In **premises** where the main area of entertainment is a hall where **the public** will participate in such activities as dancing etc, signal lamps and alarm sounders should be provided. It may however be beneficial for a two stage alarm system to be provided. (See also paragraphs 9.6 and 9.8.)

Note: Duplicate lamps should be provided with this lighting circuit so that on the failure of one signal light the other signal light will continue to function.

Automatic fire alarm systems

9.19 In the arrangements previously described reliance is placed on the people present to discover and give warning of fire. This may not be adequate if a fire starts in a part of a **building** which is visited only occasionally, such as a store room or in **premises** which have been left unattended at night time. In these two cases an automatic fire alarm system may be of great value in reducing the time between the outbreak of fire and its discovery, although in the second case, the value will only be fully realised if the system is directly connected either to a fire brigade control room or to a central alarm station. Automatic fire alarm systems are now available in various forms and are capable of carrying out a variety of functions, which include the detection of heat in the form of an unnaturally rapid rise in ambient temperature, smoke and the products of combustion and flame. Guidance is contained in **British Standard** 5839: Part 1 and advice should be sought from specialist firms or consultants and the fire brigade as to the correct type of system for given circumstances.

9.20 If a manually operated electrical system and an automatic fire alarm system are installed in the same **building**, they should preferably be incorporated into a single integral system for the purposes of providing a general alarm of fire.

9.21 Although an automatic fire alarm system will give warning of fire it cannot itself take appropriate action to contain it. However, there are automatic detection systems which have a variety of functions such as closing down a ventilation or air-conditioning plant, opening vents for the removal of smoke from escape routes, bringing a pressurization system into operation or merely activating automatic door releases as well as the giving of an alarm.

Notes/Amendments

Chapter 10: FIRE FIGHTING EQUIPMENT

10.1 All **places of entertainment** should be provided with means for fighting fire for use by persons in the **building**, and in deciding what are the appropriate means consideration needs to be given to the nature of the materials likely to be found there. The advice of the local fire brigade should be sought in cases of doubt.

10.2 Fires are classified in accordance with **British Standard** 4547 and are defined as follows:–

Class A fires –	Fires involving ordinary combustible materials such as wood, cloth, paper.
Class B fires –	Fires involving flammable liquids or liquefiable solids (petrol, paraffin, paints), oils, greases and fats.
Class C fires –	Fires involving gases.
Class D fires –	Fires involving burning metals.

Recommended standards

Class A fires

10.3 Class A fires are the most likely type of fire to occur in the majority of **premises**. Water is the most effective agent for extinguishing these fires and the appropriate equipment will therefore be hydraulic hose reels, water type extinguishers or extinguishers containing fluoroprotein foam (FP), aqueous film forming foam (AFFF) or film forming fluoroprotein foam (FFFP).

Hose reels

10.4 If hose reels are installed they should be located where they are conspicuous and always accessible, eg in corridors. Hose reels should be sited with care to avoid having to be taken through **fire doors** which need to be kept closed for safety. The hose should comply with Type 1 hose specified in **British Standard** 3169 and hose reel installations should conform with **British Standard** 5306: Part 1 and **British Standard** 5274.

Portable fire extinguishers (water or foam types)

10.5 If portable fire extinguishers are installed they should be provided and allocated to comply with clause 5.2 of **British Standard** 5306: Part 3.

Class B fires

10.6 In **buildings** (or parts of **buildings**) where there is a risk of fire involving flammable liquid it will usually be appropriate to provide portable fire extinguishers of foam (including FP, AFFF and FFFP) and carbon dioxide (CO_2), halon or powder types. Table 1 of clause 5.3 of **British Standard** 5306: Part 3 gives guidance on the minimum scale of provision of various extinguishing media for dealing with a fire involving exposed surfaces of contained liquid.

Notes: 1. Care should be taken when using halon or CO_2 extinguishers as the fumes and products of combustion may be hazardous in confined spaces. Because of its ozone depleting potential the production of halon ceased on 1 January 1994 and only banked or recycled halon is now available. Existing halon extinguishers may continue to be used but halon will eventually become more difficult to obtain and managers, owners etc should consider how any halon extinguishers will be replaced.

2. The Department of Trade and Industry has produced a new leaflet "Fire Fighting Halon Phase Out: Advice on Alternatives and Guidelines for Users" which may be obtained free of charge from DTI, 151 Buckingham Palace Road, London SW1 9SS (telephone 071-215 1018).

Class C fires

10.7 No special extinguishers are made for dealing with fires involving gases because the only effective action against such fires is to stop the flow of gas by closing the valve or plugging the leak. There would be a risk of an explosion if a fire involving escaping gas were to be extinguished before the supply could be cut off.

Class D fires

10.8 None of the extinguishing agents referred to in the preceding paragraphs will deal effectively with a fire involving metals. These fires should only be dealt with using special extinguishers by personnel trained in the handling of combustible metals. Where such materials (eg aluminium, magnesium, sodium and potassium) are used or stored it is recommended that specialist advice is sought.

General

10.9 Fire extinguishers should conform to **British Standard** 5423 and be installed and maintained as outlined in **British Standard** 5306: Part 3. Schemes for ensuring conformity with these standards have been produced by the British Standards Institution and adopted by British Approvals for Fire Equipment (BAFE) and conforming equipment and services are recognised by that organisation's mark of approval.

Fire blankets

10.10 Fire blankets are classified in **British Standard** 6575 and are described as follows:–

(a) *light duty:* These are suitable for dealing with small fires in containers of cooking fat or oils and fires in clothing;

(b) *heavy duty:* These are for industrial use where there is a need for the blanket to resist penetration by molten metals.

Electrical equipment

10.11 For fires involving electrical equipment, portable fire extinguishers of vaporising liquid gas are appropriate.

Fire fighting equipment for places of entertainment

10.12 **Premises** licensed to accommodate less than 500 persons should be provided with 1 portable water type extinguisher (rating 13A) for every 200m^2 of floor area or part thereof with not less than 2 such extinguishers on any floor. (See also paragraph 10.3.)

10.13 Manual firefighting equipment should also be provided for special risks in accordance with the following scale:–

(a) Stage exceeding 56m ^{2}:

hydraulic hose reels or 2 water type extinguishers (rating 13A on each), on each side of the **stage**, and 1 light duty fire blanket;

(b) Stage not exceeding 56m ^{2}:

1 water type extinguisher (rating 13A), on each side of the **stage**, and 1 light duty fire blanket;

(c) Dressing rooms:

in every block of 4 dressing rooms a minimum of 1 water type extinguisher (rating 13A), and 1 light duty fire blanket;

(d) Scenery store, stage basement, property store, bandroom:

1 water type extinguisher (rating 13A) in each risk area, or an appropriate extinguisher where water is unsuitable for the fire risk presented;

(e) Electrical intake rooms, stage switchboards and battery rooms:

1 carbon dioxide extinguisher or 1 dry powder extinguisher or 1 halon extinguisher (minimum rating 21B);

(f) Boiler rooms:

(i) solid fuel fired – 1 water type extinguisher (rating 13A)

(ii) oil fired – 1 dry powder or foam extinguisher (rating 34B).

Note: A water type extinguisher may be replaced by any of the types of extinguishers described in paragraph 10.3.

TYPES OF FIRE EXTINGUISHER

TYPE	COLOUR CODE	EXTINGUISHING ACTION	CLASS OF FIRE	METHOD OF USE
WATER	RED 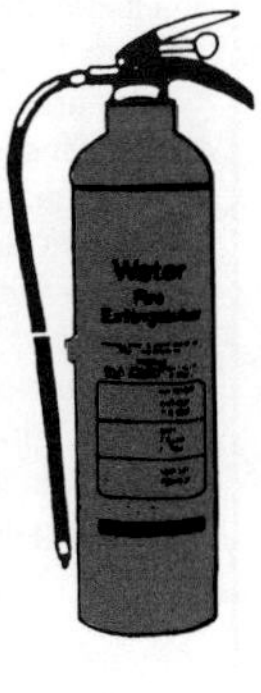	Mainly by cooling the burning material.	Class A. DANGER Do not use on live electrical equipment or on burning oil.	The jet should be directed at the base of the flames and kept moving across the area of the fire. Any hot spots should be sought out after the main fire is out.
FOAM	CREAM 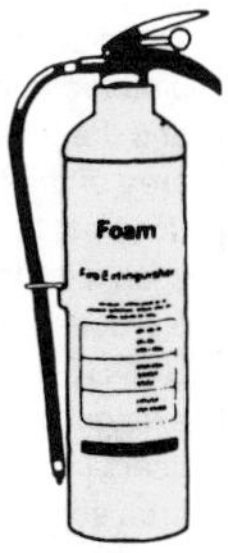	Forms a blanket of foam over the surface of the burning liquid and smothers the fire.	Class B.	The jet should not be aimed directly onto the liquid. Where the liquid on fire is in a container the jet should be directed at the edge of the container or on a nearby surface above the burning liquid. The foam should be allowed to build up so that it flows across the liquid.
AFFF (Aqueous film-forming foam) FFFP (Film-forming fluoroprotein foam) FP (Fluoroprotein foam)	CREAM 	Forms a fire extinguishing water film on the surface of the burning liquid. Has a cooling action with a wider extinguishing application than water on solid combustible materials.	Classes A and B. DANGER Some foams are not suitable for use on live electrical equipment.	For Class A fires the directions for water extinguishers should be followed. For Class B fires the directions for foam extinguishers should be followed.
DRY POWDER	BLUE 	Knocks down flames.	Class B. Safe on live electrical equipment although does not readily penetrate spaces inside equipment. A fire may re-ignite.	The discharge nozzle should be directed at the base of the flames and with a rapid sweeping motion the flame should be driven towards the far edge until the flames are out. If the extinguisher has a shut off control the air should then be allowed to clear; if the flames reappear the procedure should be repeated. WARNING Dry powder does not cool the fire very well and care should be taken to ensure that the fire does not start up again.

British Standard 5423 recommends that the bodies of water extinguishers should be coloured signal red. Other extinguishers should be predominantly red with a second colour to indicate the extinguishing medium covering an area sufficient to be readily apparent. Some manufacturers provide a colour coded label, others colour code the entire body of the extinguisher.

TYPE	COLOUR CODE	EXTINGUISHING ACTION	CLASS OF FIRE	METHOD OF USE
DRY POWDER (Multi-purpose)	BLUE 	Knocks down flames and on burning solids melts down to form a skin, smothering the fire. Has some cooling effect.	Classes A and B. Safe on live electrical equipment although does not readily penetrate spaces inside equipment. A fire may re-ignite.	The discharge nozzle should be diirected at the base of the flames and with a rapid sweeping motion the flame should be driven towards the far edge until the flames are out. If the extinguisher has a shut off control the air should then be allowed to clear; if the flames reappear the procedure should be repeated. WARNING Dry powder does not cool the fire very well and care should be taken to ensure that the fire does not start up again.
CARBON DIOXIDE (CO_2)	BLACK 	Vaporising liquid gas which smothers flames by displacement of oxygen in the air.	Class B. Safe and clean to use on live electrical equipment.	The discharge horn should be directed at the base of the flames and the jet kept moving across the area of the fire. WARNING CO_2 does not cool the fire very well; care should be taken to ensure that the fire does not start up again. DANGER Fumes from CO_2 extinguishers can be harmful to users in confined spaces. The area should therefore be ventilated as soon as the fire has been controlled.
HALON	GREEN 	Vaporising liquid gas giving rapid knock down by chemically inhibiting combustion.	Class B. Clean and light. Can also be used on small surface burning Class A fires. Effective and safe on live electrical equipment.	The gas is expelled in a jet which should not be aimed into burning liquids as this risks spreading the fire. It should therefore be aimed at the flames and kept moving across the area of the fire. WARNING Halon does not cool the fire very well and care should be taken to ensure that the fire does not start up again. DANGER Fumes from halon extinguishers can be harmful to users in confined spaces or if used on hot metal. The area should therefore be ventilated as soon as the fire has been controlled.

TYPE		EXTINGUISHING ACTION	CLASS OF FIRE	METHOD OF USE
HOSE REEL	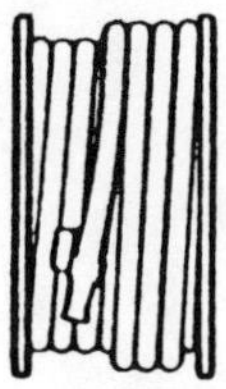	Mainly by cooling the burning material.	Class A. DANGER Do not use on live electrical equipment.	The jet should be aimed at the base of the flames and kept moving across the area of the fire.
FIRE BLANKET		Smothering.		The blanket should be placed carefully over the fire and the hands shielded from the fire. Care should be taken that the flames are not wafted towards the user or bystanders.
Light duty			Classes A and B. Suitable for burning clothing and small fires involving burning liquids.	
Heavy duty			Suitable for industrial use. Resistant to penetration by molten materials.	

Notes/Amendments

Chapter 11: **USE OF PYROTECHNICS, SPECIAL EFFECTS AND FIREARMS**

General

11.1 The purpose of this chapter is to give guidance in the appraisal of **pyrotechnics** and special effects, and to set out guidelines to follow in order to safeguard **the public**, performers and staff from injury during an entertainment.

Pyrotechnics

11.2 Application for the licensing authority's consent to the use of **pyrotechnics** and special effects should be made by the **licensee** in writing at least seven days before the first performance of the entertainment and give full details of the proposed use and the date and time on which a demonstration can be witnessed by a representative of the licensing authority.

Manufacturers' instructions

11.3 All **pyrotechnics** are potentially dangerous and should be used strictly in accordance with manufacturers' and suppliers' instructions and recommendations. In no circumstances should 'home-made' **pyrotechnics** be used as it is both dangerous and illegal to make **pyrotechnic** devices without a manufacturer's licence.

11.4 **Pyrotechnics** manufactured for use in the open air should not be used inside any **building** or temporary structure.

Competent person in charge

11.5 All **pyrotechnics** should be the responsibility of a person competent to handle them, who has been adequately trained, instructed or made aware of the correct method of control and operation, and of any special safety methods required. (See also paragraphs 11.9 to 11.25). This person should also be satisfied that any other person who may be required to use **pyrotechnics** in the course of an entertainment has been likewise instructed.

Siting of devices

11.6 Pyrotechnic devices should not be sited:–

(a) where they could discharge into an audience or orchestra pit;

(b) down-stage of a safety curtain or main tabs (except with specific **approval**);

(c) where any entrance or exit from a set or any escape route from a **stage** could be affected;

(d) where persons would be exposed to danger; and

(e) where adjacent **scenery**, **props** or drapes would be exposed to danger of ignition.

11.7 When an assessment has been made of the risks involved, a decision should be made as to where **the public**, performers and other staff should be sited in relation to the device. Care should also be exercised in respect of **scenery** and clothing particularly where productions involve the use of flimsy, or highly flammable costumes, and artificial hair wigs; or where doubt exists as to any of the materials used in the **stage** construction. The following additional precautions should be taken:–

(a) careful attention should be paid to the height reached by any flaming or hot particles, the amount and density of smoke produced, noise levels and any proposed special features;

(b) caution should be exercised where **pyrotechnics** are intended to be used in older **buildings** as they may create air movement which can cause dust and rubbish to be dislodged from the fly galleries and grid. If any doubt exists these areas should be inspected before **approval** is given; and

(c) before granting a licence allowing for the use of **pyrotechnics** the licensing authority should assess the risks involved in relation to the type of entertainment and the **premises** concerned. If they are in any doubt about public safety it is open to the licensing authority to apply more stringent conditions or refuse the application.

Firing devices

Construction of firing boxes

11.8 Pyrotechnic devices should always be fired using a properly constructed firing box (see diagram 14). This may be powered from mains electricity or from a suitable battery. Typical circuit diagrams are given in the Association of British Theatre Technicians Code of Practice for Pyrotechnics and Smoke Effects.

(a) **Mains powered**

(i) Although **pyrotechnic** devices will fire using direct mains voltage this practice is unnecessarily hazardous. A transformer should therefore be used to reduce the voltage to a safe level. Although devices will fire reliably at voltages above 24V, it is recommended that a voltage in excess of 50V be used to allow for the drop in long cables and possible high resistance connections;

(ii) The most suitable transformer is a safety isolating transformer (to **British Standard** 3535). This standard specifies output voltages not exceeding 110V. Where the output voltage exceeds 55V a centre tapping is specified which should be earthed to limit the voltage to Earth. This ensures that no dangerous voltages are present in the devices or the leads to them and also that an adequate current is available to fire the device from a 'domestic' (13 amp) power supply. The transformer should be adequately rated to supply the necessary current for firing and should be protected against overload;

(iii) The mains supply to the transformer should be controlled by a double pole key switch which retains the key unless in the 'off' position. An indicator light should indicate when the box is energised. The firing circuits should be controlled by double pole switches biased to the 'off' position, and be connected to the **pyrotechnic** device by an insulated lead and either a plug and socket system or by shrouded screw or spring terminals.

(b) **Battery powered**

(i) **Pyrotechnic** devices can be fired using battery power provided that the battery is of a suitable voltage (at least 24V and preferably 50V) and can supply sufficient current;

(ii) The battery may either be separate from the firing box and connected to it with a lead or be an integral part of the firing box. In the former case the lead should be fitted with a plug so that the battery can be isolated from the firing box. In the latter case the firing device should be fitted with a double pole key switch in the battery supply circuit, which retains the key unless in the 'off' position. A key switch should also be fitted if the separate battery is not easily portable. An indicator light should indicate when the battery power is connected and a battery condition indicator is useful;

(iii) The firing circuits should be controlled by double pole switches biased in the 'off' position, and be connected by a plug and socket system or by shrouded screw or spring terminals.

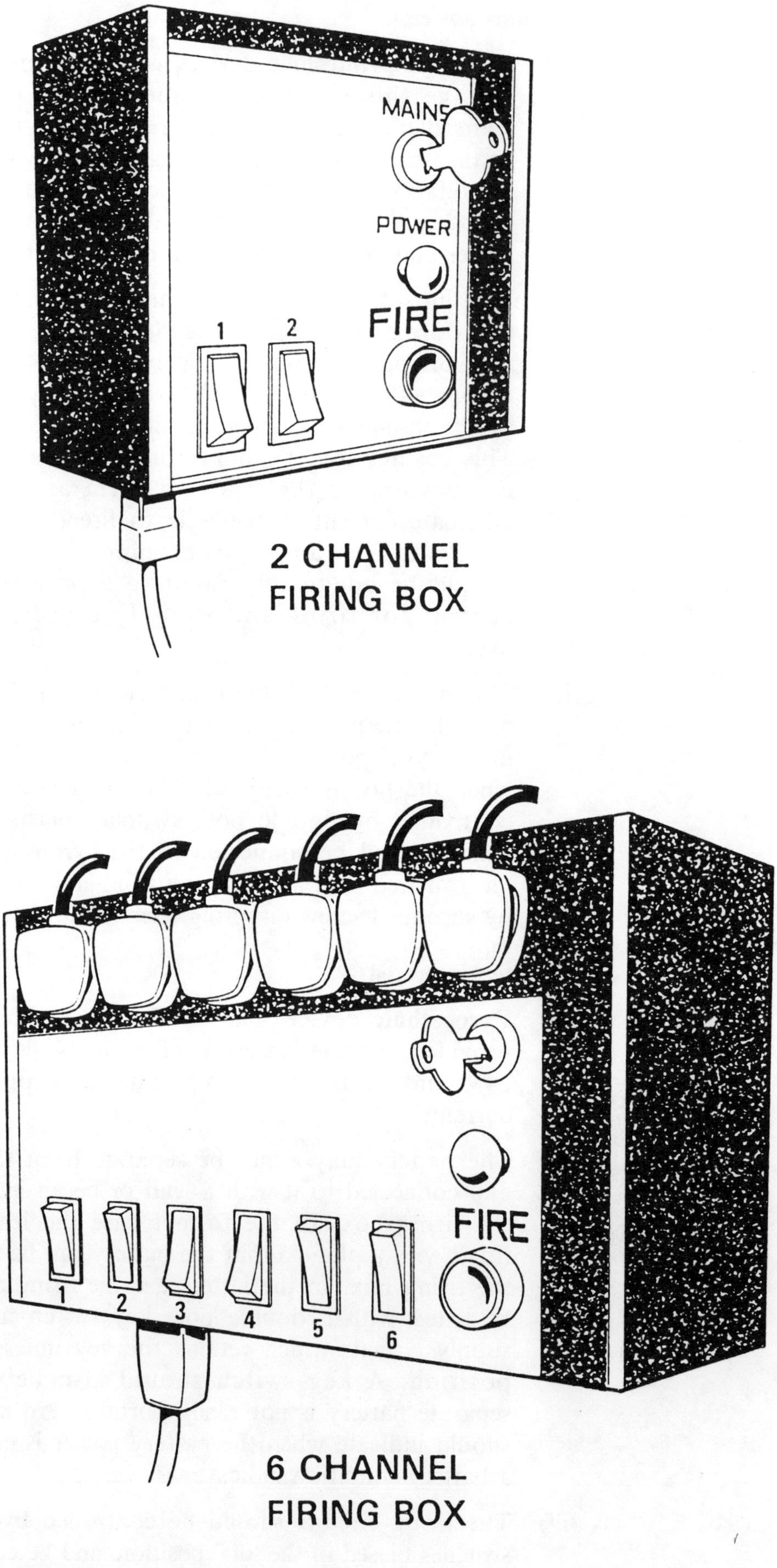

Diagram 14 Examples of firing boxes.

Safety factors

11.9 The firing device should be electrically and mechanically safe and maintained in a good condition.

11.10 All devices should be fired by the use of a firing box which should be capable of being isolated by means of a key operated isolation switch. The key should be in the possession of the operator firing the device. Electrically operated devices should be capable of being isolated from their firing supply by the complete disconnection of the supply cable.

11.11 The electrical supply for firing boxes should be constant, and not subject to fluctuation, eg reduced voltage through use of dimmers.

11.12 The following safety conditions should also be observed:–

(a) before loading any **pyrotechnic** device all firing circuits should be in the 'off' position; the mains switch should be switched off and the key withdrawn (or battery pack removed). When loading a **pyrotechnic** device it is essential that the flash box or other device is isolated by unplugging a plug and socket situated 1.5m from the flash box. After loading the device the operator(s) should turn their back to the device before reconnecting this local isolator. When the effects are loaded the danger area around the devices should be cleared of people and combustible material;

(b) the device should not be fired until everyone in the immediate vicinity is in a safe position;

(c) if some time is to elapse before firing, the key (or battery) should be withdrawn and kept by the authorised person until required for final checking;

(d) the firing box should not be energised until immediately prior to firing;

(e) the operator(s) should have a clear view of all effects and their danger areas and should only fire on cue if it is safe to do so;

(f) if the device misfires, or appears to misfire the following action should be taken:–

(i) all firing circuits should be switched to 'off';

(ii) the device should be disconnected at the mains supply or the battery pack removed;

(iii) the lead should be disconnected from the firing box and the local plug and socket by the device unplugged;

(iv) the device should be left for at least 15 minutes;

(v) the **pyrotechnic** should be removed from the position of firing and immediately placed in a non-combustible container, covered with dry sand, and removed to open air in a non-public area;

Note: The device should NOT be placed in water.

(vi) when **the public** have left the **premises** the system should be reconnected and an attempt made to fire the defective unit. If the **pyrotechnic** should still fail to fire, the procedure in (i) to (v) should be followed and the manufacturer contacted for advice quoting the batch/code number.

*Notes: 1. A **pyrotechnic** device which appears to have misfired may nevertheless have ignited and may fire without warning, even though it has been disconnected from the firing circuit. It is therefore essential that appropriate precautions are taken before approaching and handling the device.*

*2. **Pyrotechnics** containing their own oxidising agents may continue to burn even when totally immersed in water. Metallic powders such as magnesium may burn more violently or explode if water or CO_2 is applied to them. As it will be virtually impossible to extinguish any **pyrotechnic** device once it has become ignited, the safe storage and use of these items is of paramount importance.*

11.13 The **licensee** should ensure that the person(s) carrying out the loading and/or firing action are competent to do so. An untrained person should not be allowed to perform this function.

Use of bomb tanks

11.14 Purpose built firing boxes, either mains or battery powered, should be used when firing maroons in bomb tanks (see diagrams 15 and 16). Maroons should be sited in the centre of the tank.

(a) **Construction**

The tank should be of all metal construction, the walls and bottom to be not less than 2.5mm thick, with all joints welded. It should have an open top (to allow for the venting of hot gases in an upward direction only). The open top should be covered with a strong wire mesh lid with a matrix of approximately 12.5mm capable of being fixed shut when in use. The dimensions of the tank should be such that the maroon can be suspended at least 200mm clear of any internal face. This size of tank is suitable for firing single maroons only (see diagram 15), although a stand-by maroon can be provided within the same tank.

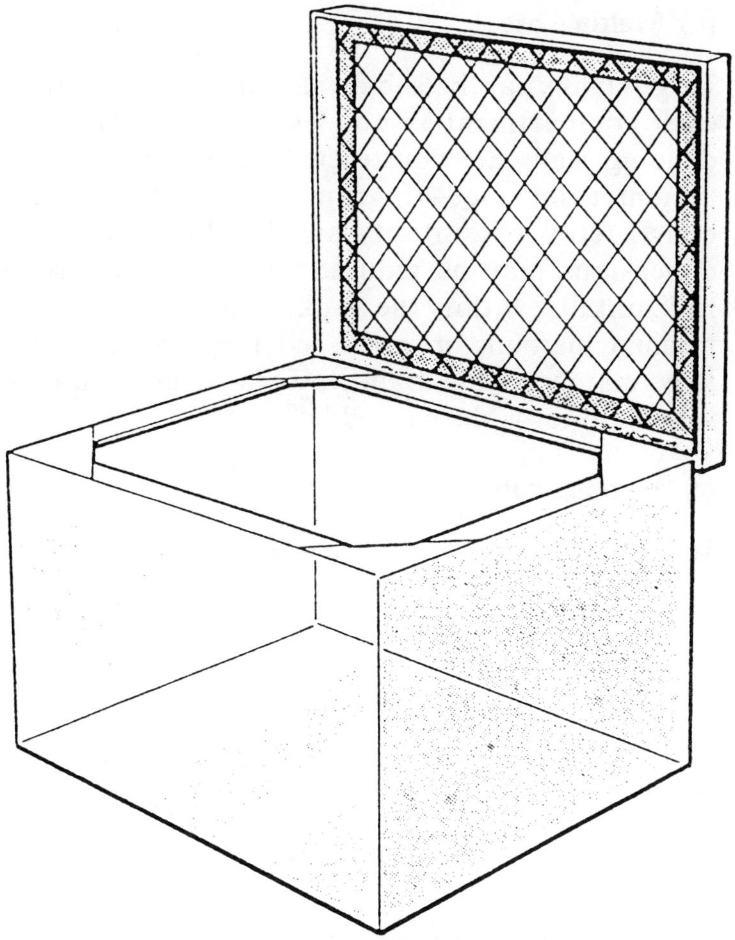

Diagram 15 Example of bomb tank.

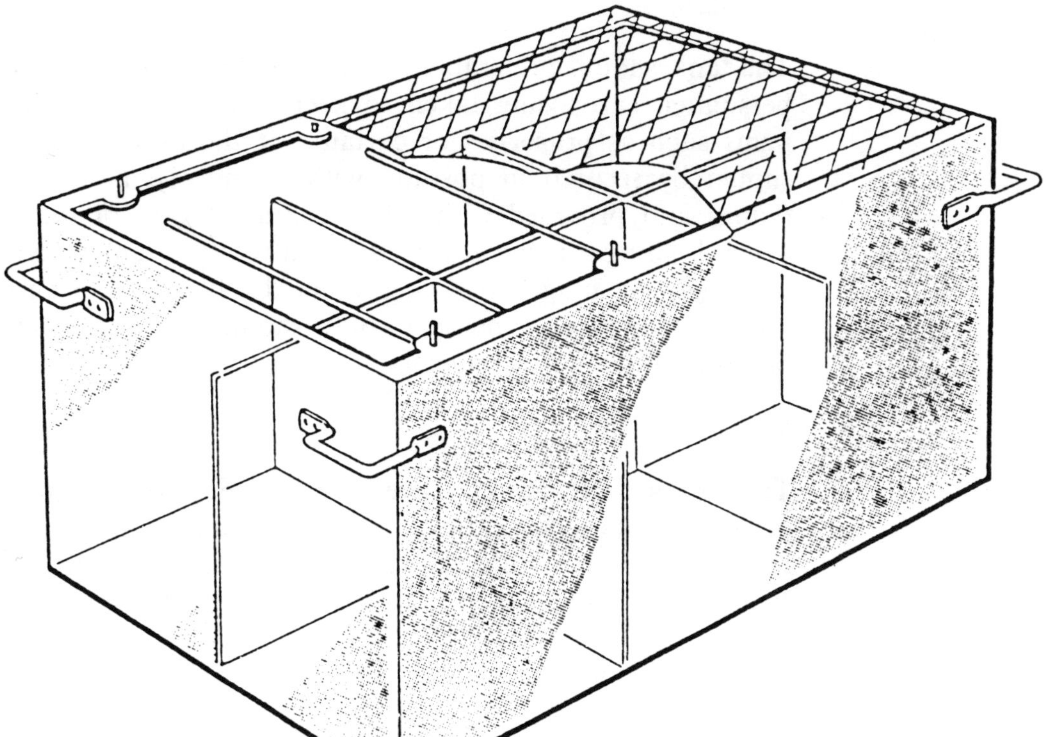

Diagram 16 Example of a multiple bomb tank.

(b) **Multiple use**

Larger tanks can be sub-divided for multiple use, provided that each compartment is not less than a 400mm cube and the dividing walls are constructed of welded metal not less than 2.5mm thick (see diagram 16). If the compartments are constructed of suitably sized cold drawn steel tubes the dimensions may be reduced. These tubes may be positioned horizontally within the tank, provided that only small and medium maroons are used and they are fired singly. However large maroons should always be used in vertical compartments, venting towards the meshed lid.

(c) **Warning notices**

When bomb tanks are in use warning notices which read

DANGER

EXPLOSIVES

KEEP CLEAR

should be positioned on the tank so that they are visible from all sides. These notices should be in block letters and not less than 50mm in height.

Where a tank can be positioned so that it can only be approached by certain defined routes, notices which read 'NO ENTRY-DANGER-EXPLOSIVES' (see diagram 17) should be placed at strategic points so that the area is effectively isolated from personnel.

(d) **Siting**

The tank should preferably be sited off the **stage** and clear of all personnel, care being taken to keep it clear of any flammable items such as **properties** or curtains. If the tank is suspended it should be secured in position with substantial non-combustible fixings and provided with safety chains in case of their failure. When loading a maroon it is essential that the tank is isolated by unplugging a plug and socket situated 2m from the tank. After loading the maroon the operator(s) should turn their back to the bomb tank before reconnecting this local isolator.

(e) **Firing**

The operator(s) should have a direct view of the **pyrotechnic** device and its surrounding area before firing unless the tank is located in a controlled area to which access is not possible. Before reloading a bomb tank, all debris should be removed.

Diagram 17 Example of warning notices.

Storage

Legislation

11.15 The Explosives Acts 1875 and 1923 govern the use and storage of explosive substances. The term 'explosive' in these Acts applies to any substance used or manufactured with a view to producing a practical effect by explosion or a pyrotechnic effect.

Storage of explosives

11.16 2 kg or more of **fireworks** must be kept in a licensed store or be stored in registered premises. Storage of over 25 kg of explosives requires annual registration by the local authority and inspecting officers will check on the quantities kept and ensure that storage arrangements are adequate. In addition the person in charge of the explosives may have to be certificated by the chief officer of police of the relevant police district as a suitable person to keep explosives.

11.17 Pyrotechnic devices are not normally stored on the **premises** but are obtained for selected events in such quantities as are required for immediate use. Where they are stored on the **premises**, provision should be made to house the storage receptacle in a suitable room or enclosure.

Storage receptacles

11.18 Receptacles should be constructed as follows:–

(a) they should be of a substantial construction, as defined in the Explosives Act 1875, eg a japanned or tinned iron or steel trunk or box, or a wooden box; and

(b) any metalwork exposed within the interior should be either non-ferrous or covered with non-ferrous metal, not readily ignitable material or paint of 1mm thickness.

11.19 A notice (with block letters not less than 25mm high) reading 'DANGER - NO SMOKING - NO NAKED FLAME' should be positioned on the outside of the lid of the receptacle.

Storage enclosure

11.20 The enclosure should be of **fire-resisting construction** and should preferably be positioned against an outside wall and provided with a vent direct to open air. It should be lockable, away from public areas and under strict management control.

11.21 It should contain no combustibles or other substances which are not compatible with the explosives being stored.

11.22 There should be no appliances with exposed flames or heating elements within the enclosure.

11.23 A 'NO SMOKING' notice should be fixed on the external face of the enclosure door and a 9 litre water type extinguisher (rating 13A) should be hung on the wall immediately outside the enclosure door.

Separate storage

11.24 **Fireworks**, maroons and flash powder should be stored separately from percussion caps, safety cartridges, safety fuse and detonators. However, if the aggregate weight does not exceed 7 kg, one receptacle could be suitable, provided it is sub-divided into compartments, each compartment being fire separated from the others.

Access

11.25 Both receptacles and enclosures should be kept locked shut at all times other than when necessary to gain access to the interior, and the keys should be kept under the control of the person responsible for the safe storage of such material.

Smoke and vapour effects

11.26 There are 2 main ways of producing smoke and vapour for special effects:–

(a) solid carbon dioxide (dry ice) fog machines

(b) fluid-based smoke machines

Note: *Smoke can also be created by the use of* ***pyrotechnics*** *and the provisions for their use are contained in earlier parts of this chapter. However, the special considerations which apply to smoke machines should also be applied to* ***pyrotechnics*** *used in this way.*

Smoke effects

11.27 Smoke machines work on the principle of superheating an oil or water-based chemical, atomising the fluid and forcing the resulting smoke out of the front of the machine under pressure (see diagram 18). The principal application of smoke machines is to produce smoke to enhance lighting effects in discos and clubs as well as special effects in theatres. Smoke machines should only be operated in accordance with the manufacturer's written instructions.

Vapour effects

11.28 To achieve an illusion of a 'dream-like' quality in theatrical productions solid carbon dioxide (dry ice) is immersed in hot water or steam (see diagram 19) so that clouds of white mist are produced which drift along at ground level. The resulting mist is frequently directed by the use of ducting and fans.

PRINCIPLE OF OPERATION OF A TYPICAL SMOKE MACHINE

Oil is injected into the coil under pressure, vaporised by the heating element and partial combustion occurs. With the absence of sufficient oxygen for full combustion only smoke is produced

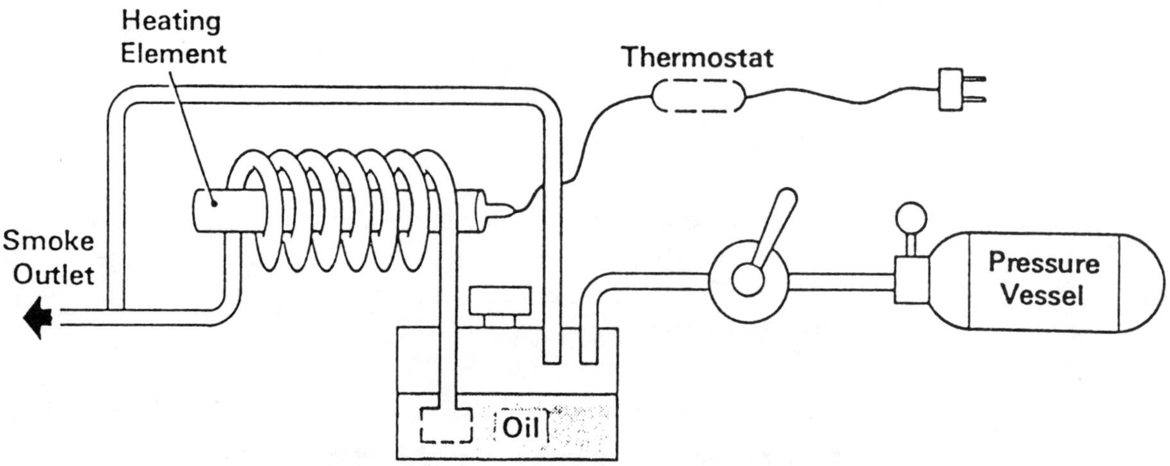

Diagram 18 Example of the principle of operation of a typical smoke machine.

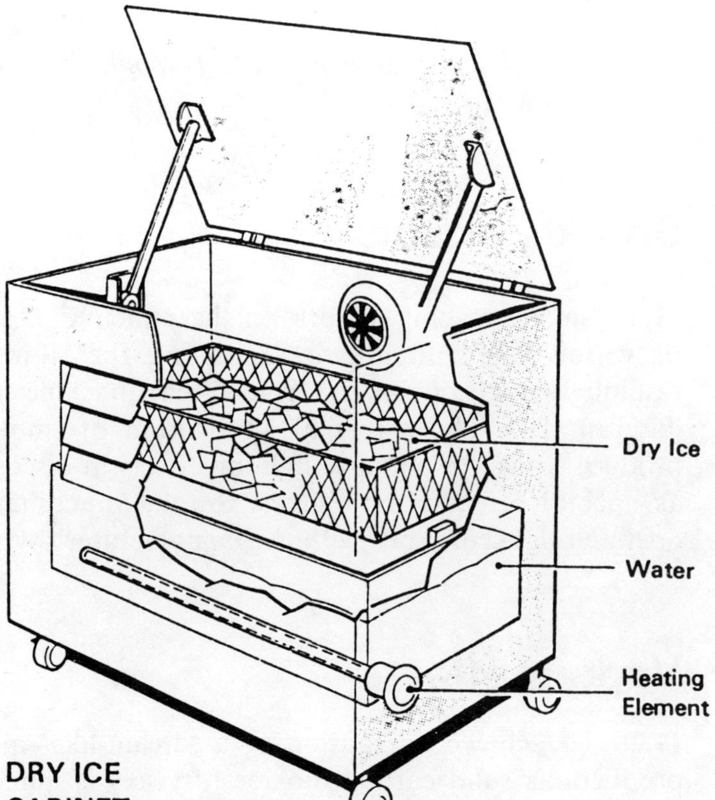

Diagram 19 Example of dry ice cabinet.

Storage and use

11.29 Dry ice should be handled only with imperforate gloves having good thermal insulation, as skin contact can cause severe burns. Medium-term storage is possible in containers with good insulation, but these should be vented and sited in well-ventilated areas. As small pieces of dry ice vaporise rapidly, blocks of dry ice should not be broken up until immediately before use.

Special considerations

Carbon dioxide

11.30 Carbon dioxide is an asphyxiant and high concentrations can present a life risk to the audience, performers and **stage** staff. Good ventilation is therefore a prerequisite. Particular care needs to be taken in **premises** which have under-stage and basement workshop/storage areas as the gas may flow into these areas through openings and crevices putting the people in these areas at risk.

11.31 Following initial generation, the vapours become invisible and the concentration of gas may be difficult to determine. If there is any doubt about the concentration present expert advice should be requested to monitor oxygen and carbon dioxide levels prior to its use at a performance.

Nitrogen

11.32 Increasingly nitrogen is being used to create vapour effects but owing to its inert characteristics specialist advice should be sought to monitor oxygen levels and to take into account the effect of convection currents during the presence of an audience.

General considerations

11.33 The smoke machine should be sited so that it does not obstruct exit routes nor cause a hazard to surrounding items such as curtains or drapes. It should also be located so that the volume of smoke/vapour is so controlled that it does not seriously affect **means of escape** or obscure exit signs. It is advisable to limit the amount of smoke/vapour in the areas to which **the public** are admitted, to the minimum necessary for the desired effect.

11.34 The machine should be in a fixed position, adequately protected against unauthorised interference, and manned by or be readily accessible to a competent operator at all times when the machine is being used. The smoke/vapour outlets should also be within sight of the operator at all times.

11.35 The **licensee** may be required to produce documentary evidence of the non-toxicity and non-flammability of the smoke or vapour produced.

11.36 Smoke/vapour outlets should not be sited where their output can be discharged or drift into exits, exitways, stairways, **means of escape** routes, etc.

11.37 Actual operation of the machine should take account of public safety and should be limited to the minimum amount of time required to achieve the necessary level of density.

11.38 Fans should be used, where necessary, to direct the smoke/vapour into the desired area, in order to prevent clouding at the point of discharge and possible overspill into other parts of the **premises**.

Note: Before the use of a smoke or vapour effect is ***approved*** *consideration should be given to any automatic fire detection or fire sensor system installed in the* ***premises****.*

Real flame

11.39 Real flame should not be used in an entertainment unless **approved**.

Lasers and strobes

11.40 The laser is a light amplifier which is capable of generating an intense parallel beam of coherent light. Those used in entertainment and display can range in power levels from a few milliwatts to tens of watts of light power. The most commonly used lasers are argon gas filled and emit a blue/green light in a very narrow beam. Krypton gas is also used to give further colours and by combining them a full range of colours can be obtained. The main hazard from the laser beam is to the eye as a highly concentrated spot is focused on the retina where it can cause damage by burning. This hazard is present in all but the very lowest powered lasers and even reflections from polished surfaces may be hazardous with medium and higher powered installations. In addition there are hazards due to high voltage electricity and the cooling water required to operate the equipment.

11.41 Lasers are becoming an increasingly popular effect but if incorrectly installed can cause damage to eyes (see paragraph 11.40 above) and certain types of lasers can present a fire risk. **Licensees** should therefore ensure that they are installed and operated by experts.

11.42 The requirements for the safe use of lasers are detailed in the Health and Safety Executive's Guidance Note No. PM19 'Use of lasers for display purposes', ISBN 0 11 883370 7, available from HMSO and **British Standard** 4803: Part 3. These standards should be adopted. The HSE Note imposes conditions to safeguard the audience and workers at the venue following the setting up of a laser installation.

11.43 Strobes are normally found in discos or at pop concerts. As the flashing lights can have a disorientating effect and have been known to induce epilepsy the **licensee** should ensure that they are installed by experts, and to reduce the hazards it is recommended that all strobe lights should operate on a fixed rate of not more than 8 flashes per second. Where more than one strobe light is in use, the flashes should be synchronised.

11.44 Where strobes are in use in a theatrical production the **licensee** should provide a warning in the foyer and/or programme to this effect.

Firearms

General

11.45 Firearms should be under the control of a responsible person, usually the holder of the firearm certificate. Only this person or one under his direct control can use firearms in **premises** which are licensed for entertainment. When not in use firearms should be stored away from the parts of the **premises** to which **the public** have access and secured against access by persons other than the certificate holder.

Use of firearms

11.46 Operators should follow this safety code:–

(a) the firearm should not be of a larger calibre than is necessary for the event;

(b) each firearm should be examined to ensure that it is free of corrosion and has been adequately maintained;

(c) the venting direction of the firearm should be examined and identified and the person using the firearm should be made aware of the venting direction;

(d) venting should not be towards either the operator or other persons on **stage**;

(e) ammunition should be checked for corrosion and crimped cartridges used wherever possible; and

(f) firearms should never be directly aimed at either the audience or at any person on **stage**.

Note: Wadded charges should not be allowed when crimped safety cartridges are obtainable for the firearm. Where wadded charges are ***permitted*** *the discharge should be directed into a safe area away from* ***the public****, performers and staff.*

Notes/Amendments

Chapter 12: **FIRE PRECAUTIONS IN SMALL PREMISES**

General

12.1 This chapter applies to small **premises** including community centres and village halls where:–

(a) the area of the hall, room or auditorium accommodation is $200m^2$ or less; and

(b) the **premises** are used by 300 members of **the public** or less; and

(c) no permanent provision is made for a closely seated audience.

Premises containing a **stage** providing entertainment in a small single storey hall are dealt with in this part of the guide.

12.2 Because of the wide variety of **premises** which fall into the categories listed in paragraph 12.1 above there is a need for a greater degree of flexibility in deciding what fire precautions are appropriate for the **premises** so that the requirements are not unduly onerous. However, the **distances of travel** etc set out under Stages 1, 2 and 3 of Chapter 5 will still apply.

12.3 The recommendations in paragraphs 12.14 to 12.20 may not be appropriate if the accommodation for **the public**:–

(a) is at ground floor level and flanks a street or an adequate dispersal space;

(b) is not used to accommodate more than 160 persons;

(c) has no basement; or

(d) if it has a basement, this is not used for the storage of combustible materials and is not an **area of high fire risk**.

12.4 It is vital that all **premises** used by **the public** are as safe as possible from fire and that **managers** know how to protect **the public** and their staff from injury or death if a fire breaks out. To this end the **licensee** should consult the licensing authority well in advance of the proposed event to ensure that his proposals and the fire precautions for the **premises** are adequate for the event(s) in question. The **licensee** should fully understand and comply with the conditions of the licence. **Managers** of non-licensed **premises** should also have an evacuation plan and should consult the fire authority for advice whenever they see a need to do so. (Further information and advice on management responsibilities is given in Chapter 1 and on **means of escape** including basic principles in paragraphs 5.2 to 5.7 of Chapter 5.)

The manager's responsibilities

12.5 The safety of **the public** is the responsibility of the **manager.** This means that the **manager** should be in charge of and present on the **premises** during the whole time that they are open to **the public**, and kept free from work which would prevent him from being immediately available in the event of an alarm of fire. (See also Chapter 1 paragraphs 1.5 and 1.7 to 1.11.)

Premises managed by voluntary organisations

12.6 In the case of **premises** which are managed by voluntary organisations and which are hired out from time to time to other organisations for public entertainments, but do not have paid staff to supervise such functions, it may not be appropriate for the **licensee,** staff or committee members to be present at every entertainment. Nevertheless a person should be nominated as responsible for both the entertainment and safety arrangements. It is also recommended that the hirer fulfills the following conditions:–

(a) is aged 18 years of age or over;

(b) signs a written undertaking to accept responsibility for being in charge of and on the **premises** at all times when **the public** are present and for ensuring that all conditions of the entertainment licence relating to management and supervision are met;

(c) possesses a full copy of the licence conditions for the **premises**;

(d) ensures that he receives instruction and training in the items listed in paragraph 1.9;

(e) carries out a safety check as listed in paragraphs 1.5 and 1.7; and

(f) provides such attendants as may be required under paragraphs 12.7-12.10; and instructs them as to their essential responsibil ities in the event of fire or other emergency.

Furthermore all voluntary management committees should be encouraged to provide instruction for their committee members and booking secretaries in emergency procedures at least once every year.

Attendants

12.7 There should be competent attendants on duty during the whole time that **the public** are on the **premises.** These attendants should have been specifically instructed as to their essential responsibilities in the event of fire or other emergency. Account should be taken of the additional responsibility caused by the attendance of disabled persons. See also Chapter 2.

12.8 Attendants should be readily identifiable to **the public** by means of some conspicuous clothing or marking system which is visible under all lighting conditions. They should carry an effective hand torch in **premises** where part or all of **the public** areas may be darkened when **the public** are present.

12.9 The number of attendants on duty on the **premises** to assist persons entering or leaving should be not less than 1 for every 250, or part of 250 persons present; and

(a) if the number of persons on the floor or tier is less than 100, there should be at least 1 additional attendant on duty on that floor or tier;

(b) if the number of persons present on any floor or tier exceeds 100 there should be at least 2 additional attendants on duty on that floor or tier.

12.10 Where most of the audience is under the age of 16, the number of attendants on duty should be not less than 1 for every 100 or part of 100 within the auditorium and 1 attendant for every 50 or part of 50 above the lowest floor.

12.11 The main duty of attendants is to ensure that safe conditions are maintained in the **premises** and to achieve this they should:–

(a) ensure that no overcrowding occurs in any part of the **premises;**

(b) keep all gangways and exits clear at all times;

(c) prevent standing on seats or furniture; and

(d) be aware of any special requirements needed to ensure the safe evacuation of the audience.

12.12 The **manager** should ensure that attendants are trained in fire procedures with the fire authority being consulted as appropriate. Instruction should include a demonstration of the locations and use of the firefighting equipment provided, how to call the fire brigade and evacuation procedures. (See also paragraphs 1.9 to 1.11, Chapters 2 and 10.)

Calling the fire brigade

12.13 *The fire brigade should be called immediately to every outbreak of fire* from the nearest available telephone. Instructions as to the location of the nearest exchange telephone and the method of calling the fire brigade should be prominently displayed within the **premises**.

Construction and fire separation

12.14 Each floor, beam, column or hanger and each load-bearing wall or other load-bearing member should be capable of resisting the action of fire for not less than 30 minutes.

12.15 Walls, partitions, roofs, floors and stairways should be constructed of **non-combustible materials** or have walls/partitions comprising materials which are not easily ignited. Any partition, wall or other form of separation which is required to provide fire separation between parts of the **premises** should, except where otherwise specified, be of **fire-resisting construction.** In the case of single floor **premises,** a roof constructed of steel trusses with wood purlin and timber boarding underdrawn with **non-combustible** materials or such similar construction may be **accepted.**

12.16 Any **platform** or similar structure should be constructed of such material and in such manner as the licensing authority may consider necessary; and any proscenium wall should be constructed of **non-combustible materials**.

12.17 Any material for acoustic or thermal insulation purposes and used as a wall or ceiling lining, or as a suspended ceiling, should be **approved** and should be fixed and supported in an **acceptable** manner.

Note: For this purpose the licensing authority will normally require such material to be ***non-combustible*** *or to be rated Class 1 of the classification for surface spread of flame when tested in accordance with the appropriate provision of* ***British Standard*** *476: Part 7, and be throughout its thickness of no greater flammability than its surface. Certain other materials may, however, be* ***accepted*** *subject to compliance with such conditions as the licensing authority may consider to be appropriate.*

12.18 Openings in walls and partitions which are required to provide fire separation (or openings in fire-resisting enclosures) should be fitted with **fire doors.**

12.19 The parts of **premises** referred to below should be separated by **fire-resisting construction:**–

(a) any cloakroom, room for staff and/or performers and any store-room;

(b) any kitchen or servery;

(c) any room or cupboard, shaft or duct which contains or constitutes a high fire risk.

12.20 Each **means of escape** (including any corridor, passageway, stairway or lobby) should be enclosed and separated from the remainder of the **premises**; and each enclosure should be of **fire-resisting construction.**

Furniture, furnishings, floor coverings and synthetic materials

12.21 Information on furniture, furnishings, floor coverings, curtains, synthetic materials and decorations is contained in Chapter 4.

Exits, stairways, doors etc

12.22 Exits should be provided and so arranged as to afford ready **means of escape** from all parts of the **premises.**

12.23 At least two separate **means of escape** should be provided from every floor and the width of each **means of escape** should be not less than the width set out in column 2 of Table E (paragraph 12.24) for the numbers given in column 1. The **means of escape** should deliver into the streets, passageways or open space, from which there should be adequate arrangements for dispersal.

12.24 The number of exits and their width should normally be calculated using the method described in paragraph 5.26. However in the case of those **premises** which are defined in sub-paragraphs (a) to (d) of paragraph 12.3 the simplified approach given in Table E below will be considered to be **acceptable.** The widths should be the distances measured between the finished wall surface at any point or, in the case of doorways, the distance between the leaves of the doors, when open at right angles to the frames.

Table E Minimum width of exits

Column 1	Column 2	Column 3
Number of persons not exceeding	Two exits each not less than	One exit not less than (see Note 1 for where **permitted**)
50	750mm	750mm
100	750mm	Not applicable
200	1.05m	"
300	1.5m	"

Notes: 1. A single exit is ***permitted*** *only in the case of* ***premises*** *at ground floor level where:–*

(a) there is adequate frontage and a ***final exit*** *to a street, passageway, or, open space;*

(b) the overall fire risk in the ***premises*** *is low;*

(c) the ***distance of travel*** *to the* ***final exit*** *is not greater than the maximum* ***permitted*** *for escape in one direction (See Table B of paragraph 5.10);*

(d) the wall and ceiling linings, curtains, drapes etc comply with the standards detailed in this guide;

(e) the door is constructed to open outwards; and

(f) a maximum of 50 persons are ***permitted*** *for the function.*

2. *Where wheelchair users may be present no exit should be less than 900mm wide.*

12.25 Where any stairway, lobby, corridor or passageway forms part of the **means of escape** from the **premises** it should be of a uniform width not less than the width of the exit, or the aggregate widths of the exits, from which it leads and should be formed and arranged so that nothing will impede the free flow of traffic or prejudice the safety of persons using it. (See also Chapter 5.)

Stairways, corridors, ramps, barriers

12.26 Information on stairways, corridors, ramps, barriers etc is contained in Chapter 5.

Floors and tiers

12.27 No floor or tier whether temporary or permanent should be constructed with a slope of more than 35 degrees to the horizontal.

12.28 The floors of gangways, lobbies, corridors and passageways and the treads of steps and stairways, should have non-slippery and even surfaces.

Doors and gates

12.29 Unless otherwise **permitted,** exit doors and gates should open in the line of exit and should be free from fastenings other than panic bolts or such fastenings as may be **approved.**

12.30 Exit doors and gates should be hung so as not to obstruct, when open, any gangway, passageway, stairway or landing. Any exit door or gate which may be required to be kept open while **the public** are on the **premises** should be provided with **approved** fastenings to keep it open in such a manner that it will not form an obstruction.

Note: Any door required to provide fire separation should not be fitted with cabin hooks or other devices for holding it in the open position.

12.31 No means should be provided for holding a self-closing **fire door** in an open position, other than by an automatic fail safe door release and this should only be fitted to a door which cannot be kept closed because of the problems presented by its frequent use in the day-to-day running of the **premises**. No such devices should be fitted to doors forming part of a stairway enclosure.

12.32 Any door or gate which opens into a corridor, stairway, passageway or gangway should, where possible, be recessed. If this is not possible it should be so hung as to be closed by a stream of persons passing through such corridor, stairway, passageway or gangway in the direction of the exit. Any such door or gate should be rendered self-closing, and should be arranged so as not to interfere with exit or other doors. Consideration should be given for such a door or gate to be provided with a vision panel.

Note: Where, for convenience of working, a door not providing fire separation is required to be held open into an exitway, it should be fitted with a holding device which will allow it to be shut readily by gentle pressure an the door itself.

12.33 A door or gate should not open immediately onto a step or steps, and a landing of not less than lm should be provided between the door or gate and the step or steps. Where a door or gate is at the head or foot of steps, a landing of not less than lm should be provided between the door or gate and the top step or the bottom step as the case may be.

12.34 Door handles or any other door fitting should preferably not project more than 50mm into exitways. Handles should be of the lever type but with the free end turned in to prevent it from catching in clothing.

12.35 Unless otherwise **permitted,** any inwards-opening door or gate, collapsible gate, roller shutter or revolving door should not be provided in or across **means of escape** from the **premises.**

12.36 Any door or gate **permitted** to open inwards should be so arranged that:–

(a) it can be locked back in such a manner as to require a key to release it; and

(b) when locked back it does not form an obstruction and does not reduce the width of the exitway.

The key should also be removed from the lock and kept in a safe place. However, (a) and (b) need not apply in the case of exit doors used only by staff where safety is not prejudiced.

12.37 Revolving doors are not permissible for **means of escape** and where they are installed they should have side-hinged doors opening in the direction of the exit immediately adjacent to the revolving doors.

12.38 Panic bolts or latches should be without sharp projections and should be of such pattern that normal horizontal pressure on the push-bar will open the door. The push-bar should be placed at a height within the range of 900mm – 1.1m from the floor. The words 'PUSH BAR TO OPEN' should be displayed conspicuously in block letters not less than 20mm, and preferably 50mm in height immediately above or below the push bar. Panic bolts should be of a strong type and the working parts should be properly maintained. As an alternative to panic bolts in smaller **premises**, push pads may be **accepted**. Panic bolts and panic latches fitted after the publication of this guide should comply with **British Standard** 5725: Part 1.

Exit and directional signs

12.39 All exit doors considered necessary for **means of escape** should be clearly and conspicuously indicated.

12.40 All fire safety signs, notices and graphic symbols should conform with **British Standard** 5499: Part 1, and with **British Standard** 2560 for internally illuminated exit signs. **Self luminous fire safety signs** should also comply with the requirements of **British Standard** 5499: Part 2. Existing signs and notices which do not meet the **British Standard** need not be replaced immediately if they are fulfilling their purpose effectively. They should however be examined and replaced if they are found to be inadequate.

12.41 Self luminous fire safety signs will not generally be allowed where the viewing distance exceeds 24m.

12.42 Maintained internally illuminated exit signs will be required where the lighting may be dimmed or extinguished.

12.43 Exit signs should be illuminated whenever **the public** are present.

12.44 A notice bearing the word 'EXIT' in plain block lettering not less than 125mm high should be placed over any door or opening leading from the place of assembly to an exit. The notice should be positioned, wherever possible, between 2m and 2.5m above the floor. No other notice should be incorporated in, or form part of, an 'EXIT' notice other than a directional arrow.

12.45 Any exit which is not a normal route of travel from the **premises** should be indicated by a notice bearing the words 'Fire exit', in conspicuous lettering of appropriate size. The notice should, wherever possible, be displayed immediately above the exit opening. Where this is not possible a position should be chosen where the notice can be seen and it is least likely to be obstructed.

12.46 Where an exit cannot be seen or where persons escaping from the **premises** may be in doubt as to the location of the exit, 'Fire exit' signs, to include a directional arrow, should be provided at suitable points along an escape route. Such notices should be fixed in conspicuous positions and wherever possible be positioned between 2m and 2.5m above the floor level.

Notices on doors

12.47 A notice with the words 'Push bar to open' should be permanently displayed immediately above the push-bar on all doors fitted with a panic bolt or panic latch. (See paragraph 12.38.)

12.48 A notice with the words 'Fire door – Keep shut' should be permanently displayed at about eye level on both faces of all **fire doors** except those to cupboards. **Fire doors** which are normally open but which close automatically on the operation of fire detectors should bear the words 'Automatic fire door – Keep clear'.

12.49 A notice with the words 'Fire door – Keep locked shut' should be permanently displayed on the outside face of all **fire doors** not required to be self closing eg cupboards.

12.50 A notice with the words 'Slide to open' and with an arrow pointing in the appropriate direction should be permanently displayed at about eye level on the face(s) of all sliding doors which afford an exit or are situated across an escape route.

12.51 A notice with the words 'Fire escape – Keep clear' should be permanently displayed at about eye level on the external face of all doors which are provided solely as a **means of escape** in case of fire and which, because they are not normally used, are liable to be obstructed.

Note: All fire notices should be in conspicuous lettering and of appropriate size.

Electrical, gas and oil installations

12.52 All electrical, gas and oil installations should comply with the appropriate recommendations contained in Chapter 6 of this guide.

Lighting

12.53 All parts of the **premises** to which **the public** have access and all external exitways should be provided with **normal lighting** capable of providing sufficient illumination of those parts for **the public** to leave the **premises** safely. (See also paragraphs 6.29 and 6.30.)

12.54 A minimum standard of illumination should be maintained so that the **building** cannot be plunged into total darkness whether deliberately or accidentally. **Emergency lighting** installed to comply with **British Standard** 5266 will be deemed to satisfy this requirement. (See also paragraph 6.35.) The **emergency lighting** system should be regularly serviced and should be tested before **the public** are admitted to the **premises.**

*Notes: 1. In premises used occasionally for audiences up to 100 people, **emergency lighting** may be provided by torches. Such torches, which should be of adequate strength to guide people out of the **premises** safely, should be issued to attendants. Additional torches should be provided in dressing rooms and back stage areas for performers to use. If **emergency lighting** is provided by this means an adequate supply of spare batteries should be available.*

*2. Another form of **emergency lighting** (which may satisfy the enforcing authority as meeting the requirements of paragraph 6.35) is battery and lamp units. These are often suitable for occasional use and may be hired for the particular event.*

12.55 If at any time there is a failure of the lighting referred to in paragraph 12.53 above, and that lighting is either by gas or electricity, the parts of the **premises** affected, including the notices referred to in paragraph 12.43 should immediately be illuminated by some alternative means which enable **the public** to see their way out of the **premises** in safety.

12.56 The **normal lighting** should be kept on at all times when **the public** are on the **premises**, except in any part which is well-lit by daylight, and should illuminate all exit fire safety signs unless they are **self-luminous fire safety signs.**

Heating

12.57 Any stove, electric or gas fire or **heating appliance** other than low pressure hot water pipes and radiators should be protected by guards. The taps of gas fires should be of a design to avoid malicious or accidental interference. Portable heaters should be securely fixed to the wall at a safe height. Portable liquefied petroleum gas appliances should not be kept on the **premises** while **the public** are present. (See also paragraph 6.58.)

Seating and gangways

12.58 The type of seating arrangements adopted in **places of entertainment** will vary with the use to which the **premises** are put.

Note: *For guidance on **premises** where there is permanent provision for closely seated audiences see Chapter 7.*

12.59 In circumstances where more than 250 temporary seats are to be used in the auditorium the following arrangements should apply:–

(a) chairs or other single seats should be secured together in lengths of not fewer than 4 seats and not more than 12 so that the seats cannot be separated from each other merely by pushing one or more seats in the row (the only exception being chairs in boxes or other **approved** enclosures); and

(b) provision should be made for the rows of seating flanking the gangways to be fixed to the floor effectively preventing the individual seats or rows of seats from being dislodged into the gangways or from being toppled over, except that:–

(i) only end seats of such rows need to be fixed to the floor if all seats in each row are secured together; or

(ii) only the end seats of each length of seating referred to in paragraph 12.59(a) above which form such rows need be fixed to the floor.

Note: *In **premises** which are intended to be used only occasionally for closely seated audiences, where the fixing of seating to the floor is impractical or undesirable (eg on polished dance* *floors) the use of floor bars instead of floor screws may be* ***permitted.*** *Such floor bars should be no more than 25mm in height, have a cambered top surface so as to avoid the risk of tripping persons using the seatways, and should extend from the row to be fixed to at least two adjacent rows, but should not extend across any gangways. (See also Chapter 7.)*

12.60 Seats should be of adequate strength, soundly formed and free from defects which could cause injury to **the public.**

12.61 Where seating for more than 100 people is provided, a seating layout should accompany the application for a licence. A scale plan of the arrangements for seats, gangways and exits should be readily available for inspection.

12.62 Seating and gangways in an auditorium or hall should be so arranged as to allow free and ready access direct to the exits of these areas, with no portion of any gangway normally more than 18m from an exit measured along the line of the gangway.

12.63 In enclosures containing seating, displays etc, gangways should be arranged and maintained free of obstruction so that reasonably direct access to the exits is provided.

12.64 Gangways should be of sufficient width for the occupancy. Their width should be calculated in accordance with the width of exits, but should in no case be less than 1.05m wide. The creation of bottle-neck conditions which could cause obstruction to persons making their way out of the **premises** should be avoided.

12.65 Where steps are used as gangways, the risers should be sealed or formed so that people cannot inadvertently step through them. Stair treads should have a going of not less than 280mm exclusive of any nosing. Risers should be not more than 190mm in height. The width, rise and going of steps should be uniform. Handrails should be provided as necessary.

Stage and stage area

12.66 All **scenery** including cloths, draperies, gauze-cloths, floral decorations, hangings, curtains and all fabric decorations on the **stage** should be flame retarded and should be so maintained. **Scenery** or **properties** should not be kept or used:–

(a) in the **stage** basement except when required for use in a current production; or

(b) in any part of the **premises** other than on the **stage** or in an **approved property** or **scenery store.**

12.67 Lighted candles should not be used for decorative purposes. Suitable protection for lamps such as lampshades made from not readily ignitable materials should be provided for lighting. Other appliances which may become hot should be set well apart from **scenery** or any other combustible material.

12.68 Separation should, where appropriate, be formed between the audience and the **stage** by provision of panels or curtains of heavy woollen or flame-retardant treated materials.

12.69 The **platform** or **stage** should not be congested with **scenery** or **properties** and the exits leading from the **stage** should be maintained free from obstruction.

12.70 Further information on **stages, scenery** and **properties** is contained in Chapter 8.

Warning in case of fire

12.71 Information on fire alarms and automatic fire alarm systems is contained in Chapter 9 and paragraphs 9.3 and 9.4 are particularly appropriate.

First aid fire fighting equipment

12.72 Information on fire fighting equipment is contained in Chapter 10.

Notes/Amendents

Chapter 13: SPECIAL PROVISIONS FOR TEMPORARY STRUCTURES AND PLACES OF ENTERTAINMENT WHICH ARE UNDER COVER IN OTHERWISE OPEN AIR SITUATIONS

Additional definitions applicable to this chapter

Air supported structure means a **structure** that has a space-enclosing single skin membrane anchored to the ground and kept in tension by internal air pressure so that it can support applied loading. The membrane may or may not be restrained by trans-surface ropes or cables.

Distance of travel means the actual distance that a person must travel between any point in the **structure** and a **final exit.**

Final exit means an exit from a **structure** into open air from which persons can disperse quickly and safely.

Pneumatic structure means a **structure,** or a substantial part thereof, which depends upon air pressure to maintain shape and includes:–

(a) a double wall enclosing membrane, or

(b) a flexible frame giving support to the enclosing membrane, which in each case are inflated to provide shape and stability to the **structure.**

Room includes the area enclosed within a **structure** or, where the enclosed area is physically divided by barriers of any kind, any area so separated from the remainder of the **structure.**

Structure(s) includes a **tent,** a **pneumatic structure** or an **air supported structure.**

Temporary structure means a **structure** which is designed, constructed and erected in such a way that it may be moved from the site either whole or after being dismantled.

Tent means a **structure** (including a marquee), the covering of which is mainly of flexible material which achieves its sole support by beams, columns, poles, arches, ropes and/or cables, or mechanical means.

General

13.1 It is vital that all **structures** used by **the public** are as safe as possible from fire and that **managers** know how to protect **the public** and their staff from injury or death if a fire breaks out. To this end the **licensee** should consult the licensing authority well in advance of the proposed event so that there is no misunderstanding about the use of the **structure** or about the nature of any fire precautions which are regarded as necessary. The need for this early consultation is even more important where a **temporary structure** will be erected just prior to the event as it is sometimes difficult to establish exactly what circumstances will prevail when these facilities are in use. This guidance is given in order to avoid last minute embarrassment to organisers who might be required to carry out extensive alterations, or even to cancel their event at short notice because of inadequate fire safety arrangements. Where there is no requirement to obtain a licence it is nevertheless recommended that **managers** should consult the fire authority for advice.

Site selection and access

13.2 The following fire safety factors should be considered by the **licensee** when choosing a site:–

(a) access and egress for **the public,** emergency vehicles and equipment;

(b) the proximity of surrounding **buildings** and vegetation and other fire risks in relation to the spread of fire;

(c) the need for a telephone (to call the fire brigade);

(d) availability of mains services; and

(e) the slope or unevenness of the ground.

13.3 An outline site plan of all **structures** should be prepared by the **licensee** showing the position of all entrances and exits, generator equipment, vehicles, etc. It should be kept up to date on the site and be readily available for inspection. The plan should be agreed by the licensing authority following consultation with the fire authority having regard to occupancy, use, position and other factors relevant to safety. It should not be altered without reference to the licensing authority.

13.4 The site should be arranged so as to allow for adequate means of access by firefighting appliances to within 50 metres of any part of the **structure.** Access routes should be not less than 4 metres wide, should have no overhead **structure** or cable less than 4.5 metres above the ground, and should be capable of taking the weight (about 12.5 tonnes) of firefighting appliances in all weathers. Emergency vehicle routes within the site should be kept clear of obstruction at all times.

13.5 Access to hydrants and other water supplies should not be obstructed or obscured.

Spacing between temporary structures, caravans and other structures

13.6 In order to reduce the possibility of the spread of fire between **structures, temporary structures** and caravans, the spacing should normally be not less than 6 metres unless compensating fire safety features are provided.

The manager's responsibilities

13.7 The safety of **the public** is the responsibility of the **manager.** This means that the **manager** should be in charge of and present in the **structure** during the whole time that the **structure(s)** are open to **the public**, and kept free from work which would prevent him from being immediately available in the event of an alarm of fire. (See also Chapter 1 paragraphs 1.4 to 1.5 and 1.7 to 1.11).

13.8 Before **the public** is admitted to any performance or event in the **structure** the **manager** should ensure that a safety check is carried out on the lines described in paragraph 1.5 and that the requirements of paragraph 1.7 are met.

Attendants

13.9 There should be competent attendants on duty during the whole time that **the public** are in the **structure.** These attendants should have been specifically instructed as to their essential responsibilities in the event of fire or other emergency. Account should be taken of the additional responsibility caused by the attendance of disabled persons. See also Chapter 2.

13.10 At least one attendant who should not be the person in charge of the **structure** should be on duty for every 250 or part of 250 persons in the **structure**, except that, where most of the patrons are under the age of 16, there should be at least one attendant for every 75 or part of 75 persons.

Note: The licensing authority may decide that because of the type of entertainment and/or composition of the audience additional attendants should be on duty (see paragraph 1.4).

13.11 Attendants should be readily identifiable by means of some conspicuous clothing or marking system which is visible under all lighting conditions.

13.12 The main duty of attendants is to ensure that safe conditions are maintained in the **structure** and to achieve this they should:–

(a) ensure that no overcrowding occurs in any part of the **structure;**

(b) keep all gangways and exits clear at all times;

(c) prevent standing on seats or furniture; and

(d) be aware of any special requirements needed to ensure the safe evacuation of the audience/patrons.

13.13 The **manager** should ensure that attendants are trained in fire procedures with the fire authority being consulted as appropriate. Instruction should include a demonstration of the locations and use of firefighting equipment provided, how to call the fire brigade and evacuation procedures. (See also paragraphs 1.9 to 1.11, Chapters 2 and 10.)

Smoking

13.14 As a general rule smoking should not be allowed anywhere in the **structure.**

Calling the fire brigade

13.15 *The fire brigade should be called immediately to every outbreak of fire* from the nearest available telephone. Instructions as to the location of the nearest exchange telephone and the method of calling the fire brigade should be prominently displayed. If an exchange telephone is not readily available, consideration should be given to the provision of portable telephone facilities.

Construction

General

13.16 From the safety point of view, **tents,** marquees, **pneumatic structures**, **air supported structures** and similar **structures** will present difficulties not generally found in conventional **buildings**. These will range from managing the provision and maintenance of the structural and other safety features, to difficulties for **the public** who will often find their routes of egress to be over uneven ground, temporary flooring, duckboards, ramps, stairways, etc rather than the permanent surfaces to be found in conventional **buildings**. Consideration therefore needs to be given to the varying activities and the type and scale of the assembled audience.

13.17 In addition to the normal ignitability and flame spread hazard associated with the use of large areas of textile materials, it is possible that other hazards, ie high levels of toxicity, could also be present through the use of certain modern materials, eg fluorocarbon polymers, in the construction of the **structure.** The nature of all the materials used in the **structure** should therefore be established and advice sought if unusual materials are present.

Tented structures

13.18 New manufactured membranes and fabrics should be of **inherently flame-retarded fabric** or **durably flame-retarded fabric** when tested to **British Standard** 5438 Tests 2A and 2B with a 10 second flame application time in each case. Existing materials and **structures** manufactured to the previous **British Standard** 3120 continue to be **acceptable.** Other sheet materials should be Class 1 surface spread of flame in accordance with **British Standard** 476: Part 7. Materials should be free of flaming molten droplet characteristics and should not readily support combustion.

Notes: 1. *With regard to existing tented **structures** the standard quoted above (**British Standard** 3120) is only considered to be **acceptable** during the intervening period of the normal replacement of old stock. During the same replacement period of old stock licensing authorities may also wish to consider whether non-**flame-retarded fabric**, ie fabric which does not meet either **British Standard,** is **acceptable** for certain ancillary uses.*

2. *It should also be noted that the British Standards Institution have recently published **British Standard** 7157. This method of test is also **acceptable.***

13.19 The licensing authority may require the **licensee** to provide a certificate or other evidence from a body of recognised standing to show compliance with the provisions of paragraph 13.18.

13.20 All supporting poles, frames, guys, stakes, anchors, fastenings etc should be regularly tested by the supplier and maintained in a safe condition. The **structure** should be erected by a competent person and should not readily collapse when exposed to fire.

Air supported and pneumatic structures

13.21 The recommendations of **British Standard** 6661 apply only to **air supported structures**. The materials used in the construction of **pneumatic structures** should therefore comply with the recommendations contained in paragraph 13.18. However, for the purposes of this guide, where the **structure** is to be used for licensed entertainment, the recommendations contained in paragraphs 13.34 to 13.76 of this guide should be followed in preference to clauses 20, 21 and 22 of **British Standard** 6661.

Air supported structures

13.22 The **structures**, including foundations and all other elements necessary for the safety of occupants, should be inspected at regular intervals and should be adequately maintained.

13.23 The **structures** should have adequate emergency exits, **emergency lighting**, emergency exit signs, maximum and minimum pressure limiting devices and audible warnings, automatically operated standby fans and an alternative power source to operate under pressure loss and/or failure of the primary power source and an emergency support system.

Pneumatic structures

13.24 A risk peculiar to **pneumatic structures** is that their stability is dependent upon a supply of air under pressure which is provided and maintained by mechanical fans. Additionally such **structures** comprise extensive areas of flexible membrane material which, like those used for **tents**, requires particular care in selection.

13.25 For these reasons, safety measures peculiar to **pneumatic structures** include reliability of air supply systems supplemented by secondary support systems to maintain clear exit routes in case of collapse together with the selection of suitable (flame-retarded) membrane materials. The **distance of travel** from any part of the **structure** to the protection of such a secondary support system should not be more than 12 metres, with a minimum headroom of 2.5 metres.

Performance in fire

13.26 The membrane of an **air supported structure** should not readily support combustion. Experience has shown that PVC coated polyesters and polyamides and unreinforced polythene generally perform satisfactorily under fire conditions. Details of fire tests which have been carried out are given in paragraph 13.27.

13.27 A small number of fire tests on **air supported structures** of various sizes have shown that:-

(a) a fire produces smoke and toxic fumes which fill the **structure** if the membrane is not damaged. The quantity of fumes and the subsequent smoke level depend on the nature of the fire, but a small **structure** will fill with smoke more quickly than a large one. The opening of exits has the effect of causing smoke movement towards them. High level vents reduce this effect; and

(b) if the flames in a fire reach and damage the membranes at a high level (above 2.5m), other gases will be driven towards the hole so produced thus decreasing the density of smoke and the hazard to the occupants. The internal pressure will, however, be reduced and, depending on the size of the hole, the fan capacity and the number of doors opened, the **structure** may collapse.

Fire spread

13.28 Rigid linings such as wall boards used in **structures** should be not less than a Class 1 surface of flame rating as described in **British Standard** 476: Part 7.

Furniture, furnishings, floor coverings etc

13.29 Information on furniture, floor coverings, synthetic materials and decorations is contained in Chapter 4.

Means of escape for enclosed or covered structures

General

13.30 In the event of a fire breaking out in a **structure** it is essential that everyone is made aware of the danger and is able to reach a **place of safety** before being overcome by smoke, toxic gas or other products of combustion.

13.31 On becoming aware of the need to evacuate the **structure** persons will frequently look first to the way they entered but if this cannot be reached (perhaps because of the position of the fire or smoke) an understandable instinct will be to turn away from the danger and seek an alternative route to a **place of safety**. **The public** may nevertheless underestimate the risk and may be reluctant to use exits with which they are unfamiliar. The role of training staff in ensuring that **the public** react and leave promptly should not be understated.

13.32 Persons should be able to walk to safety along a clearly recognisable route by their own unaided efforts regardless of where the fire might break out in the **structure**. However, it is recognised that it will be difficult, if not impossible, for some people with disabilities to make their way to a **place of safety** without the assistance of others. (See also Chapter 2.)

Occupant capacity

13.33 The technical requirements for **means of escape** from **buildings** (including **structures**) which are common today and the factors governing these requirements have developed through the course of this century from research, observations and experience in many countries. These factors are largely interdependent and include evacuation time, the distance to exits, and the number, width and siting of exits. Collectively, the number and width of exits determine the 'exit capacity' which together with the siting of exits determines the number of persons allowed – the 'occupant capacity'. See also paragraphs 5.16 to 5.20.

Exits

13.34 Every **structure** should be provided with exits which are sufficient for the number of occupants in relation to their width, number and siting. Normally no exit should be less than 1.05m wide. It should be noted that the width between two side poles in a tented **structure** is about 950 mm.

13.35 The required minimum width of any **final exit** in relation to the number of occupants who may have to use it should be determined by Table F below:–

Table F Minimum width of exits

Minimum clear width of exit	Number of persons
1.05 metres	1 to 160
1.65 metres	161 to 240
1.95 metres	241 to 320

*Note: The maximum **acceptable** value of width of exit is 4 units. (See also Chapter 5, Table D of paragraph 5.26 and Summary, and paragraph 3.4)*

13.36 Where the **room** is intended to hold more than 50 persons there should be not less than 2 exits (see paragraph 13.39)

13.37 Where the **room** is intended to hold more than 50 persons, an allowance should be made for the fact that one exit may be obstructed and not available because of fire or smoke or other hazard. All exits should comply with the widths required by Table F of paragraph 13.35.

13.38 Any additional exit necessary to comply with the provisions of paragraphs 13.36 and 13.37 may discharge through an adjacent **room**, provided that there is a clear and direct path to a **final exit**.

13.39 The required **final exits** should be distributed evenly around the **structure** so that genuine alternative routes are available from all parts.

13.40 From any part of a **structure** having more than one exit the **distance of travel** should normally be not more than 18 metres. After the first 6.5 metres, the remainder of the route should lead in a different direction to the route of any alternative exit. In **rooms** having only one exit, the **distance of travel** should be not more than 6.5 metres.

Note: In relation to circus "big tops" it may not always be possible to restrict the ***distance of travel*** *to 18 metres. In these circumstances an assessment should be made as to the probable time it would take a person to leave any point within the* ***structure*** *to reach a* ***place of safety*** *and whether other suitable fire safety provisions are in place. Factors which should be taken into account are as follows:—*

(a) the numbers occupying the structure;
(b) the siting and number of exits;
(c) the seating arrangements and other internal layout;
(d) the number of trained staff on duty at any time;
(e) the means of calling the fire brigade in the event of an emergency;
(f) the means for giving warning in case of fire;
(g) signposting; and
(h) the provision of emergency lighting.

13.41 In determining the **distance of travel** an extra factor should be added where there are temporary ramps or stairways along the route. If the route is to include temporary ramps or stairways, an additional 0.25 metres should be added to the calculation of **distance of travel** for every 1 metre of the ramp or stairway. Thus, if the actual distance to the **final exit** is 18 metres, of which 10 metres is a stairway, the **distance of travel** should be regarded as 20 metres.

Ramps

13.42 Guidance on ramps, handrails and protective barriers is contained in paragraphs 5.79 to 5.85.

Floors and tiers

13.43 No floor or tier whether temporary or permanent should be constructed with a slope of more than 35 degrees to the horizontal. Floor coverings should be fitted so that they do not constitute an obstruction or tripping hazard.

Exit routes

13.44 Exit routes should be kept free from obstruction at all times.

13.45 While any person is within the **structure**, the doors of the **structure** and of any **room** within the **structure** which affords a means of exit, should not be locked or fastened in such a manner that they cannot easily and immediately be opened from the inside. All doors on an exit route should open outwards so that people can escape in an emergency, without impediment. Where exit doors have to be secured against outside intruders, they should be fitted with panic bolts or latches.

13.46 Panic bolts or latches should be without sharp projections and should be of such pattern that normal horizontal pressure on the push-bar will open the door. The push-bar should be placed at a height within the range of 900mm–1.1m from the floor. The words 'PUSH BAR TO OPEN' should be displayed conspicuously in block letters not less than 20mm, and preferably 50mm in height immediately above or below the push bar. Panic bolts should be of a strong type and the working parts should be properly maintained. Panic bolts and panic latches fitted after the publication of this guide should comply with **British Standard** 5725: Part 1.

13.47 Exits which consist of wall flaps or similar material are only considered suitable for use where small numbers of persons are accommodated. In such circumstances the exits should be boldly indicated at the edges and so arranged as to be opened easily and immediately from the inside. Whenever more than 50 persons are to be accommodated in a tented **structure** all exits affording **means of escape** should consist of conventional outward opening doors fitted with suitable emergency fastenings and clearly indicated.

13.48 Guy ropes, tent pegs and stakes should not be allowed to obstruct a route to a **place of safety**. Where they flank such a route, they should be shielded to form a passageway, or marked and illuminated so they can be seen clearly at all times.

Exit and directional signs

13.49 All exits considered necessary for **means of escape** should be clearly and conspicuously indicated.

13.50 All fire safety signs, notices and graphic symbols should conform with **British Standard** 5499: Part 1, and with **British Standard** 2560 for internally illuminated exit signs. **Self-luminous fire safety signs** should also comply with the requirements of **British Standard** 5499: Part 2. Existing signs and notices which do not meet the **British Standard** need not be replaced immediately if they are effectively fulfilling their purpose. They should however be examined and replaced if they are found to be inadequate.

13.51 **Self-luminous fire safety signs** will not generally be allowed where the viewing distance exceeds 24m.

13.52 Maintained internally illuminated exit signs will be required where the lighting may be dimmed or extinguished.

13.53 Exit signs should be illuminated whenever **the public** are present.

13.54 A notice bearing the word 'EXIT' in plain block lettering not less than 125mm high should be placed over any door or opening leading from the place of assembly to an exit. The notice should be positioned, wherever possible, between 2m and 2.5m above the floor. No other notice should be incorporated in, or form part of, an 'EXIT' notice other than a directional arrow.

13.55 An exit which is not a normal route of travel from a **structure** should be indicated by a notice bearing the words 'Fire exit'. The notice should, wherever possible, be displayed immediately above the exit opening. Where this is not possible a position should be chosen where the notice can be seen and it is least likely to be obstructed.

13.56 Where an exit cannot be seen or where persons escaping from the **structure** might be in doubt as to the location of the exit, 'Fire exit' notices, to include a directional arrow, should be provided. Such notices should be fixed in conspicuous positions at suitable points along an escape route and wherever possible should be positioned between 2m and 2.5m above the floor level.

13.57 A notice with the words 'Fire escape – Keep clear' should be permanently displayed at about eye level on the external face of all doors which are provided solely as a **means of escape** in case of fire and which, because they are not normally used, are liable to be obstructed.

Electrical and gas safety

13.58 Electrical installations should be installed and maintained in accordance with the provisions of the **IEE Regulations for Electrical Installations.** In respect of gas safety the Gas Safety (Installation and Use) Regulations 1984 should be consulted. Chapter 6 provides further details.

13.59 Containers of liquefied petroleum gas (LPG) should be protected against unauthorised interference and accidental leakage. LPG cylinders, both full and empty, should be kept in safe positions in the open air away from other flammable materials, or where this is not reasonably practicable because of exceptional circumstances, in an adequately ventilated storeroom constructed in accordance with Health and Safety Executive Guidance Note CS 4 – 'The keeping of LPG in cylinders and similar containers' (HMSO, ISBN 0 11 883539 4). The storeroom should not be used for any purpose other than the storage of LPG cylinders. LPG cylinders in use should be sited outside the **structure** with their valves readily accessible in case of an emergency.

Heating

13.60 Any **heating appliance** should be of an **approved** type and should be installed to the satisfaction of the licensing authority.

13.61 Heating appliances not fixed in position or using naked flame should *not* be allowed in **structures** while **the public** are present. If used at other times, they should be under constant supervision and should be switched off or extinguished and removed to a safe place outside the **structure** when not in use.

Lighting

13.62 All parts of the **structure** to which **the public** have access and all external exitways should, if intended for use in the absence of daylight, be provided with **normal lighting** capable of providing sufficient illumination of those parts for **the public** to leave the **structure** safely. (See also paragraph 6.29 and 6.30.)

13.63 A minimum standard of illumination should be maintained so that the **structure** cannot be plunged into total darkness whether deliberately or accidentally. **Emergency lighting** installed to comply with **British Standard** 5266 will be deemed to satisfy this requirement. The **emergency lighting** system should be regularly serviced and should be tested before **the public** are admitted to the **structure**.

13.64 If at any time there is a failure of the lighting referred to in paragraph 13.62 the parts of the **structure** affected, including the signs referred to in paragraph 13.53 should immediately be illuminated by some alternative means which enable **the public** to see their way out of the **structure** in safety.

13.65 The **normal lighting** should be kept on at all times when **the public** are in the **structure**, except in any part which is well-lit by daylight, and should illuminate all exit fire safety signs unless they are **self-luminous fire safety signs.**

13.66 In the case of **temporary structures** which are erected for a limited period, or are used as part of a travelling entertainment, a current certificate signed by a competent person (see paragraph 6.2) confirming that the **emergency lighting** installation meets the requirements of **British Standard** 5266: Part 1 should be **acceptable**, subject to a functional test prior to the first admission of **the public**.

13.67 Heat producing luminaires should be installed sufficiently clear of combustible materials to avoid the risk of fire.

Seating and gangways

13.68 There should in all cases be an unobstructed seatway of at least 305mm measured between perpendiculars between the back of one seat and the front of the seat immediately behind.

13.69 The number of seats in a row should not exceed:–

(a) 7 seats where there is a gangway at one end only. However, longer rows of up to 11 seats may be allowed if the seatway is increased by 25mm for each seat over 7; and

(b) 14 seats where there is a gangway at each end. However, longer rows of up to 22 seats with a gangway at each end may be allowed if the seatway is increased by 25mm for each additional pair of seats and/or any odd seat over 14.

Notes: 1. For seating arrangements at outdoor sports grounds the Guide to Safety at Sports Grounds should be consulted (HMSO, ISBN 0–11–341001–8).

2. For seating arrangements at pop concerts and like occasions see Chapter 7 paragraphs 7.21 to 7.24.

13.70 Seats in **tents** and marquees holding more than 30 people should either be fixed securely in position or rigidly linked together in rows. No row should be less than 4 seats. Seats on a sloping floor should be fixed in position.

13.71 In circumstances where more than 250 temporary seats are to be used in the **structure** the following arrangements should apply:–

(a) chairs or other single seats should be secured together in lengths of not fewer than 4 seats and not more than 12 so that the seats cannot be separated from each other merely by pushing one or more seats in the row (the only exception being chairs in boxes or other **approved** enclosures); and

(b) provision should be made for the rows of seating flanking the gangways to be fixed to the floor effectively preventing the individual seats or rows of seats from being dislodged into the gangways or from being toppled over, except that:–

(i) only end seats of such rows need to be fixed to the floor if all seats in each row are secured together; or

(ii) only the end seats of each length of seating referred to in paragraph 13.71(a) which form such rows need be fixed to the floor.

*Note: In **structures** which are intended to be used only occasionally for closely seated audiences where the fixing of seating to the floor is impractical or undesirable the use of floor bars instead of floor screws may be **permitted.** Such floor bars should be no more than 25mm in height, have a cambered top surface so as to avoid the risk of tripping persons using the seatways, and should extend from the row to be fixed to at least two adjacent rows, but should not extend across any gangways. (See also Chapter 7.)*

13.72 Seats should be of adequate strength, soundly formed and free from defects which may cause injury to **the public**.

13.73 Where seating for more than 100 people is provided, a seating layout should accompany the application for a licence. A scale plan of the arrangements for seats, gangways and exits should be readily available for inspection.

13.74 Spaces beneath tiered seating should not be used for storage and should be made as secure as practicable. Such spaces and the spaces between and under timber flooring should be kept clear of accumulations of rubbish, and these spaces should be checked between performances.

13.75 In enclosures containing seating, displays etc, gangways should be arranged and maintained free of obstruction so that reasonably direct access to the exits is provided.

13.76 Gangways should be of sufficient width for the occupancy. Their width should be calculated in accordance with the width of exits, but should in no case be less than 1.05m wide. The creation of bottle-neck conditions which might cause obstruction to persons making their way out of the **structure** should be avoided.

13.77 Where steps are used as gangways, the risers should be sealed or formed so that people cannot inadvertently step through them. Stair treads should have a going of not less than 280mm exclusive of any nosing. Risers should not be more than 190mm in height. The width, rise and going of steps should be uniform. Handrails should be provided as necessary.

Enclosed structures having a stage

13.78 Information on **stages, platforms, stage areas, scenery** and **properties** is contained in Chapter 8.

Warning in case of fire

13.79 Means should be provided for giving warning in case of fire. The means for giving the warning should be such as to avoid creating alarm for **the public** whilst giving instant warning to appropriate members of staff. The use of a public address system with an agreed procedure may be sufficient. Information on fire alarms is contained in Chapter 9.

First aid firefighting equipment

13.80 Information on firefighting equipment is contained in Chapter 10.

Special effects

13.81 The use of any special effects, eg lasers, strobes, **pyrotechnics,** should be **approved.** (See also Chapter 11.)

Other fire precautions

Grass, vegetation and combustible material

13.82 All grass and vegetation around **structures** should be kept as short as possible and cuttings should be removed.

13.83 The spaces beneath and between vehicles, caravans and trailers should not be used for the storage of combustible material.

Rubbish disposal

13.84 Rubbish of all kinds should be deposited in non-combustible containers of a suitable type, and arrangements should be made for its removal at suitably frequent intervals.

Flammable liquids

13.85 Flammable liquids such as petrol (in quantities not exceeding the prescribed maximum) and diesel fuel should be kept in suitable closed containers and should be stored safely, secure against unauthorised interference. Containers should be marked clearly to indicate the contents.

13.86 Flammable liquids should not be stored in loose containers beneath lorries, trailers etc nor near live cables or electrical equipment. **Managers** may obtain advice with regard to the storage of these liquids from the fire authority.

13.87 Particular care should be taken when the fuel tanks of vehicles and generator sets are being filled from cans or drums. Refilling should not take place while the engine or generator is running, and should be carried out with adequate ventilation, preferably in the open air. Generators and pumps should be secured against unauthorised access.

Additional provisions for circuses and like events involving animals

13.88 Animals are liable to panic in fire and in view of this there should be additional safeguards to ensure that **the public** are able to use the **means of escape** to reach a **place of safety** without being endangered by frightened animals. To this end the following measures should be taken:–

(a) the circus site should be enclosed by an **acceptable** barrier at a distance of not less than 6m from the perimeter of the **structure** ie the circus **tent.** The enclosing barrier should be provided with openings having an aggregate clear width at least equal to the total width of all the exits from the **structure**;

(b) areas through which dangerous animals pass, in tunnels or on leads should be fenced off from **the public** and sufficient clear space maintained so as not to prejudice the **means of escape**;

(c) animals should be housed and transported in suitable enclosures and all necessary measures taken to ensure that they are not a danger to **the public**;

(d) dangerous animals should be securely housed in cages or other enclosures. All doors should have locks with independent safety chains so that they cannot be opened inadvertently or maliciously. Warning signs alerting **the**

public should be in conspicuous lettering and of appropriate size;

(e) the security of all cages and enclosures should be inspected regularly, at least once a day, by a person competent to do so, and no animals should be admitted to a cage or enclosure until the security of the installation has been checked;

(f) when animals in cages are open to **public** viewing, secure barriers at least 1m high should be erected to keep **the public** at a safe distance from the cages. Notices warning **the public** of the danger, and the need to closely supervise children, should be in conspicuous lettering and of appropriate size;

(g) when animals are housed in **tents** or similar **structures** they should be secured as necessary to ensure that in a fire situation they do not run amok. There should be sufficient attendants/keepers who can be called upon to control the animals in the event of an emergency;

(h) each cage or enclosure which is regularly assembled and dismantled (such as the tunnels to, and the cages enclosing, performing areas) should be fastened together to form a secure unit. The security of the fastening of one section to another should not depend on any single bolt or connection. When such a unit has been erected, the security of all connections and of the overall enclosure should be checked by a competent person who is experienced in the construction of such units but who has not been involved in the erection;

(i) all sections of a temporary cage or enclosure should be securely anchored to the ground to prevent animals from lifting or dislodging them;

(j) when animals are performing, the person responsible for the display should be fully trained in how to respond in a fire situation, to ensure that **the public** are safeguarded from the animals and are able to reach a **place of safety**;

(k) smoking should be prohibited in animal quarters;

(l) straw and fodder for animals, and materials of a similar nature, which are required, but are not for immediate use, should be contained in metal receptacles with properly fitting covers. Such materials should be stored in a separate enclosure which is suitably protected against interference and where smoking is prohibited;

(m) whenever wild animals are present there should be sufficient trained attendants to deal with any emergency;

(n) in no circumstances should liquefied petroleum gas (LPG) be stored near animals. (See also paragraph 13.59.)

Notes: 1. The control of dangerous animals is prescribed under the Dangerous Wild Animals Act 1976.

2. The licensing of zoos is prescribed under the Zoo Licensing Act 1981.

Notes/Amendments

Appendix A: SURFACE FINISHES OF WALLS, CEILINGS, PARTITIONS ETC

Paragraph 4.6 refers.

In the early stages of a fire in a **building**, the personal hazard to occupants can be severely affected by the surface linings and finishes of walls and ceilings, and of partitions, space dividers and similar vertical surfaces. Materials likely to be found in situ in existing **buildings** are often difficult to assess in terms of their contribution to spread of flame and the development of fires and in particular care should be exercised where cellular plastic surfaces are not protected by a rigid organic finish. Some examples of these materials have therefore been broadly classified into 8 groups with an indication as to the extent to which each group may be **acceptable** for particular locations as follows:–

A: Inorganic group

Brickwork, blockwork, concrete, plasterboard, ceramic tiles, plaster finishes (including rendering on wood or metal laths), and all other surfaces conforming to **Class 0** classification as defined in Appendix A of Approved Document B of the Building Regulations or E17 of the Building Standards (Scotland) Regulations. **Acceptable** in all locations.

B: Cellulosic group

(Not flame-retardant treated)

Timber, hardboard, particleboard (chipboard), blockboard and any other material not lower than Class 3 surface spread of flame rating. **Acceptable** only on small areas of wall surface in any room (but *not* for ceilings). The area covered should not exceed half the floor area of the room or $60m^2$, whichever is smaller. Not **acceptable** on escape routes ie stairways, corridors, entrance halls.

C: Cellulosic group

(Flame-retardant supported by evidence of treatment including painting to achieve Class 1 surface spread of flame rating)

Acceptable only on walls *and* ceilings in small rooms of floor area not exceeding $4m^2$ and in other rooms for small areas of wall surface within the maximum area **permitted** for Cellulosic Group B materials, and for ceilings. Not **acceptable** on escape routes ie stairways, corridors, entrance halls.

D: Woodwool slab

Acceptable in all locations

E: Plastics – Thermosetting

(Decorative laminates)

Acceptable as for Group B unless there is evidence that they are of flame-retardant grade (to Class 1 surface spread of flame rating) in which case acceptability will be as for Group C. Where plastic materials achieve a **Class 0** classification these will be **acceptable** in all locations.

F: Plastics – Thermoplastics

(Expanded polystyrene wall and ceiling linings)

Acceptable, provided not finished with gloss paint, on inorganic surfaces in thickness not exceeding 5mm on walls, 12mm on ceilings, in same situations as Group B. Expanded polystyrene surfaces which have been painted with gloss paint should be removed.

G: Thin vinyl and paper coverings

(Other than heavy flock wallpapers)

Acceptable in all locations on an inorganic surface.

H: Heavy flock wallpapers

Acceptable as for Group B unless there is evidence that they are of flame-retardant grade (to Class 1 surface spread of flame rating) in which case acceptability will be as for Group C.

Appendix B: USE OF BATTERIES

1. The failure of battery powered equipment to operate correctly and for the required period could have serious consequences where the equipment is provided for fire safety purposes. It is therefore important to ensure that batteries used in fire safety equipment receive regular servicing and maintenance.

Equipment using primary (non-rechargeable) batteries

2. Such equipment should have the batteries replaced at regular intervals not exceeding 6 months. Sufficient spare batteries should be kept on the **premises** to replace all batteries used in the equipment, and care should be taken to ensure the rotation of the stock and that records of battery replacement are maintained.

Equipment using secondary (rechargeable) batteries

3. Due to the wide range of batteries available for use in fire safety equipment it is beyond the scope of this guide to detail the maintenance requirements for them all. However, the following general rules should be observed:-

(a) detailed records should be kept of battery replacement and maintenance so that evidence can be produced to the licensing authority of satisfactory performance and maintenance of any equipment that is powered by batteries either continuously or in the event of failure of the normal supply;

(b) only batteries recommended by the equipment manufacturer should be used;

(c) all batteries should be replaced at the end of the design life as specified by the battery manufacturer or sooner in the event of failure to provide the rated performance; and

(d) all batteries should be used and maintained strictly in accordance with the manufacturer's instructions.

Notes: 1. *It is important, in the case of self-contained* ***emergency lighting*** *units and* ***emergency lighting*** *systems, that the routine inspection and tests detailed in* ***British Standard*** *5266 are carried out. In the case of* ***premises*** *used only occasionally the daily inspection should be carried out on those days when the* ***premises*** *are used for public entertainment and recreation.*

2. *Where batteries are provided at a single point to provide power to a number of remote units, full details of the system, type, number and capacity of the cells together with the maintenance instructions should be kept available in the near vicinity of the batteries.*

3. *Smoking should be prohibited in any rooms in which electric batteries are stored or used when the batteries are capable of discharging a flammable gas. Normally the danger is associated with lead acid type batteries when they are being recharged. The rooms should be adequately ventilated.*

4. *Notices should be displayed, stating that smoking is prohibited, in and immediately outside any rooms in which electric batteries are stored or used. Any new 'No smoking' sign should comply with* ***British Standard*** *5499: Part 1.*

Appendix C: BRITISH STANDARDS REFERENCES

References to **British Standards** in this guide are to those Standards in force at the time of publication. Where **places of entertainment** were brought into use at a time earlier Standards or versions of Standards prevailed, those Standards should normally be regarded as **acceptable**. Where **places of entertainment** are brought into use after the publication of this guide, the references should be construed as references to the latest suitable versions of the relevant Standards when the work is undertaken.

British Standards relevant to the guide are listed below:-

BS 476:– FIRE TESTS ON BUILDING MATERIALS AND STRUCTURES:

- Part 4 Non-combustibility test for materials.
- Part 6 Method of test for fire propagation for products.
- Part 7 Method for classification of the surface spread of flame of products.
- Part 8 Test methods and criteria for the fire resistance of elements of building construction.
- Part 20 Method for determination of the fire resistance of elements of construction (general principles).
- Part 21 Methods for determination of the fire resistance of loadbearing elements of construction.
- Part 22 Methods for determination of the fire resistance of non-loadbearing elements of construction.
- Part 23 Methods for determination of the contribution of components to the fire resistance of a structure.
- Part 24 Method for determination of the fire resistance of ventilation ducts.

BS 559:– SPECIFICATION FOR ELECTRIC SIGNS AND HIGH-VOLTAGE LUMINOUS-DISCHARGE-TUBE INSTALLATIONS

*BS 1892:– GYMNASIUM EQUIPMENT:

- Part 2 Specification for safety requirements for mats, mattresses and landing areas.

ØBS 2560:– SPECIFICATION FOR EXIT SIGNS (INTERNALLY ILLUMINATED)

BS 2782:– METHODS OF TESTING PLASTICS:

- Part 1 Method 140D; method 140E.
- Part 5 Method 508A.

BS 3169:– SPECIFICATION FOR FIRST AID REEL HOSES FOR FIRE FIGHTING PURPOSES

---

* To be published in 1990

Ø Currently being revised and will be issued as BS 5499: Part 3

BS 3535:–	SPECIFICATION FOR SAFETY ISOLATING TRANSFORMERS FOR INDUSTRIAL AND DOMESTIC PURPOSES
BS 4293:–	SPECIFICATION FOR RESIDUAL CURRENT-OPERATED CIRCUIT BREAKERS
BS 4422:–	GLOSSARY OF TERMS ASSOCIATED WITH FIRE
BS 4547:–	CLASSIFICATION OF FIRES
BS 4607:–	NON-METALLIC CONDUITS AND FITTINGS FOR ELECTRICAL INSTALLATIONS:
Part 1	Specification for fittings and components of insulating material.
Part 2	Rigid PVC conduits and conduit fitting. Imperial units.
BS 4678:–	CABLE TRUNKING:
Part 1	Steel surface trunking.
Part 4	Specification for cable trunking made of insulating material.
BS 4790:–	METHOD FOR DETERMINATION OF THE EFFECTS OF A SMALL SOURCE OF IGNITION ON TEXTILE FLOOR COVERINGS (HOT METAL NUT METHOD)
BS 4803:–	RADIATION SAFETY OF LASER PRODUCTS AND SYSTEMS:
Part 3	Guidance for users.
BS 5266:–	EMERGENCY LIGHTING:
Part 1	Code of practice for the emergency lighting of premises other than cinemas and certain other specified premises used for entertainment.
BS 5274:–	SPECIFICATION FOR FIRE HOSE REELS (WATER) FOR FIXED INSTALLATIONS
BS 5287:–	METHOD. ASSESSMENT AND LABELLING OF TEXTILE FLOOR COVERINGS TESTED TO BS 4790
BS 5306:–	FIRE EXTINGUISHING INSTALLATIONS AND EQUIPMENT ON PREMISES:
Part 1	Hydrant systems, hose reels and foam inlets.
Part 2	Sprinkler systems.
Part 3	Code of practice for selection, installation and maintenance of portable fire extinguishers.
BS 5410:–	CODE OF PRACTICE FOR OIL FIRING:
Part 1	Installations up to 44kW output capacity for space heating and hot water supply purposes.

	Part 2	Installations of 44kW and above output capacity for space heating, hot water and steam supply purposes.
BS 5423:–		SPECIFICATION FOR PORTABLE FIRE EXTINGUISHERS
BS 5438:–		METHODS OF TEST FOR FLAMMABILITY OF TEXTILE FABRICS WHEN SUBJECTED TO A SMALL IGNITING FLAME APPLIED TO THE FACE OR BOTTOM EDGE OF VERTICALLY ORIENTED SPECIMENS
BS 5440:–		CODE OF PRACTICE FOR FLUES AND AIR SUPPLY FOR GAS APPLIANCES OF RATED INPUT NOT EXCEEDING 60 kW (1st AND 2nd FAMILY GASES):
	Part 1	Flues.
	Part 2	Air Supply.
BS 5499:–		FIRE SAFETY SIGNS, NOTICES AND GRAPHIC SYMBOLS:
	Part 1	Specification for fire safety signs.
	Part 2	Specification for self-luminous fire safety signs.
	Part 3	See BS 2560
BS 5546:–		CODE OF PRACTICE FOR INSTALLATION OF GAS HOT WATER SUPPLIES FOR DOMESTIC PURPOSES (2nd FAMILY GASES)
BS 5588:–		FIRE PRECAUTIONS IN THE DESIGN AND CONSTRUCTION OF BUILDINGS:
	Part 8	Code of practice for means of escape for disabled people.
BS 5651:–		CLEANSING AND WETTING PROCEDURES FOR USE IN THE ASSESSMENT OF THE EFFECT OF CLEANSING AND WETTING ON THE FLAMMABILITY OF TEXTILE FABRICS AND FABRIC ASSEMBLIES
BS 5655:–		LIFTS AND SERVICE LIFTS:
	Part 1	Safety rules for the construction and installation of electric lifts.
	Part 5	Specification for dimensions of standard electric lift arrangements.
	Part 6	Code of practice for selection and installation.
	Part 7	Specification for manual control devices, indicators and additional fittings.
	Part 10	Specification for the testing and inspection of electric and hydraulic lifts.
BS 5725:–		EMERGENCY EXIT DEVICES:
	Part 1	Specification for panic bolts and panic latches mechanically operated by horizontal push-bar.
BS 5810:–		CODE OF PRACTICE FOR ACCESS FOR THE DISABLED TO BUILDINGS

BS 5839:– FIRE DETECTION AND ALARM SYSTEMS IN BUILDINGS:

Part 1 Code of practice for system design, installation and servicing.

Part 2 Specification for manual call points.

BS 5852:– FIRE TESTS FOR FURNITURE:

Part 1 Methods of test for the ignitability by smokers' materials of upholstered composites for seating.

Part 2 Methods of test for the ignitability of upholstered composites for seating by flaming sources.

BS 5864:– CODE OF PRACTICE FOR THE INSTALLATION OF GAS-FIRED DUCTED-AIR HEATERS OF RATE INPUT NOT EXCEEDING 60 kW (2nd FAMILY GASES)

BS 5867:– SPECIFICATIONS FOR FABRICS FOR CURTAINS AND DRAPES:

Part 1 General requirements.

Part 2 Flammability requirements.

BS 5871:– CODE OF PRACTICE FOR THE INSTALLATION OF GAS FIRES, CONVECTORS AND FIRE/BACK–BOILERS (2nd FAMILY GASES) AND AMEND MENT No2.

BS 6004:– SPECIFICATIONS FOR PVC-INSULATED CABLES (NON-ARMOURED) FOR ELECTRIC POWER AND LIGHTING

BS 6099:– CONDUITS FOR ELECTRICAL INSTALLATIONS:

Part 2 Particular specifications:
Section 2.2: Specification for rigid plain conduits of insulating material.

BS 6141:– SPECIFICATION FOR INSULATED CABLES AND FLEXIBLE CORDS FOR USE IN HIGH TEMPERATURE ZONES

BS 6172:– CODE OF PRACTICE FOR INSTALLATION OF DOMESTIC GAS COOKING APPLIANCES (2nd FAMILY GASES)

BS 6180:– CODE OF PRACTICE FOR PROTECTIVE BARRIERS IN AND ABOUT BUILDINGS

BS 6230:– SPECIFICATION FOR INSTALLATION OF GAS-FIRED FORCED CONVECTION AIR HEATERS FOR COMMERCIAL AND INDUSTRIAL SPACE HEATING OF RATED INPUT EXCEEDING 60 kW (2nd FAMILY GASES)

BS 6262:– CODE OF PRACTICE FOR GLAZING FOR BUILDINGS

BS 6400:– SPECIFICATION FOR INSTALLATION OF DOMESTIC GAS METERS (2nd FAMILY GASES)

BS 6575:– SPECIFICATION FOR FIRE BLANKETS

BS 6644:– SPECIFICATION FOR INSTALLATION OF GAS-FIRED HOT WATER BOILERS OF RATED INPUTS BETWEEN 60 kW AND 2 MW (2nd AND 3rd FAMILY GASES)

BS 6661:– GUIDE FOR DESIGN, CONSTRUCTION AND MAINTENANCE OF SINGLE-SKIN AIR SUPPORTED STRUCTURES

BS 6714:– SPECIFICATION FOR INSTALLATION OF DECORATIVE GAS LOG AND OTHER FUEL EFFECT APPLIANCES (1st, 2nd AND 3rd FAMILY GASES)

BS 6798:– SPECIFICATION FOR INSTALLATION OF GAS-FIRED HOT WATER BOILERS OF RATED INPUT NOT EXCEEDING 60 kW

BS 7157:– METHOD OF TEST FOR THE IGNITABILITY OF FABRICS USED IN THE CONSTRUCTION OF LARGE TENTED STRUCTURES

Copies of British Standards may be obtained from the Sales Department, British Standards Institution, Linford Wood, Milton Keynes, MK14 6LE (Telephone 0908 220022)

Appendix D: REFERENCE SOURCES

1. The 1934 Home Office Manual of Safety Requirements in Theatres and other Places of Public Entertainment.

2. Guides to the Fire Precautions Act 1971.

3. The Memorandum of Guidance to the Cinematograph (Safety) Regulations 1955.

4. The GLC 'Places of Public Entertainment Technical Regulations'.

5. HSE Code of safe practice at fairs.

6. Buckinghamshire Fire Brigade/Fire Service College Code of Practice for 'Outside Events Needing a Licence for Public Music, Singing and Dancing'

7. British Standards Specifications/Codes of Practice.

8. Regulations for Electrical Installations published by The Institution of Electrical Engineers.

9. British Gas IM/15 Manual Valves – A Guide to Selection for Industrial and Commercial Gas Installations.

10. British Gas IM/16 Guidance Notes on the Installation of Gas Pipework, Boosters and Compressors in Customer's Premises.

11. British Gas IM/21 Guidance Notes for Architects, Builders, etc on the Gas Safety (Installation and Use) Regulations 1984.

12. Guide to the Draft Cinemas (Safety) Regulations – Admission of disabled and blind people to cinemas.

13. Building Regulations Approved Documents.

14. GLC: Code of Practice – Means of Escape in Case of Fire.

15. Explosives Act 1875.

16. Petroleum (Consolidation) Act 1928 – licensing provisions.

17. Petroleum Spirit (Motor Vehicles etc) Regulations 1929 - Storage of petrol other than when licensed.

18. Highly Flammable Liquids and Liquefied Petroleum Gases Regulations 1972 and Certificate of Approval No. 1 [form 2434] – for standards of storage and use.

19. Heating Appliances (Fireguards) Regulations 1973.

20. The Local Government (Miscellaneous Provisions) Act 1982.

21. The Civic Government (Scotland) Act 1982.

22. London Government Act 1963.

23. Petroleum Spirit (Plastic Containers) Regulations 1982 – for standards and storage requirements for plastic cans.

24. Gas Safety (Installation and Use) Regulations 1984.

25. Furniture and Furnishing (Fire) (Safety) Regulations 1988

26. HSE Guidance Note CS2 – The storage of highly flammable liquids – for general storage information.

27. HSE Guidance Note CS4 – The keeping of liquefied petroleum gas (LPG) in cylinders and suitable containers.

28. HSE Guidance Note PM19 – Use of lasers for display purposes.

29. Association of British Theatre Technicians: Code of Practice for Pyrotechnics and Smoke Effects.

30. GLC/London Fire Brigade: Pyrotechnics/Special Effects.

31. Guide to Design of Cavity Barriers and Fire Stops (F Spiegelhalter)

32. Post War Building Studies No 20.

33. Notes on the completion of Form FDR1.

34. Theatres Act 1968.

35. Building Regulations/Building Standards (Scotland) Regulations.

36. GLC: Parks and Open Spaces – Circus Regulations.

37. Report on Means of Escape In Case of Fire – Building Industries National Council Advisory Committee on Building Acts and Byelaws (1945).

Notes/Amendments

INDEX

	Paragraph(s)
Access:	
– and inner rooms	5.27-5.28
– Committees (England, Wales, Scotland)	2.15
– for fire appliances	3.5
Accommodation:	
– of children	1.17-1.19, 5.9
– stairways	5.68
Aggregation of exits	5.26
Air supported structures	13.21-13.23
Alarms, fire – see fire alarms	9.1-9.21
Amusement arcade – occupant load factor	5.20
Area:	
– of high fire risk	4.1-4.2, 5.27
– where standing permitted	7.9
Artificial foliage	4.20-4.23
Assembly hall – occupant load factor	5.20
Attendants:	1.4, 1.13-1.20
– at large discotheques	1.16
– at pop concerts	1.16
– audiences under age 16	1.18
– commissionaire not acceptable as	1.20
– having responsibility for children	1.19
– identification of	1.15
– number of	1.16, 1.18, 1.20
– primary duty of	1.14
– staff not to be counted as	1.20
Automatic:	
– door releases	5.103
– fire alarm systems	9.19-9.21
– fire detection	4.1, 4.10, 5.27, 9.19
Balconies	5.53
Bar area – occupant load factor	5.20
Barriers:	
– across escape routes	5.86
– protective	5.82-5.85
Basements:	
– fire resistance of floor immediately over	4.1-4.2
– stairways to	5.66-5.67
Batteries:	
– emitting flammable gas during charging	6.22
– use of	Appendix B
Billiard room - occupant load factor	5.20
Bingo hall - occupant load factor	5.20
Blackouts, complete, in theatre auditoria	6.36
Blind people	2.10

– representatives of	2.3, 2.15
– sources of advice concerning	2.15
Bomb tanks, construction and use of	11.14
Bowling alley – occupant load factor	5.20
Boxes, seating in	7.17, 7.19
Bridges	5.53
British Standards References	Appendix C
Building:	
– enclosed within confines of a complex	3.5
– forming part of town centre redevelopment scheme	3.5
– in a pedestrian precinct	3.5
– in a piazza	3.5
– on elevated podium deck	3.5
– with only one escape stairway	5.56, 5.65
Building Standards (Scotland) Regulations	Introduction, 4.3, 4.5
Calculated occupant capacity	5.19, 7.8
Call points for fire alarms	9.5
Caravans:	
– combustible materials not to be stored beneath	13.83
– spacing between	13.6
Carbon dioxide vapour effects – safety	11.30-11.31
Cavities, fire stopping of	4.9
Ceiling, surface of	4.6, Appendix A
Cellular foam	4.24-4.27
Centre on Environment for the Handicapped	2.15
Check before public admitted	1.5
Children:	1.16-1.19
– separated from parents/guardians	1.18-1.19, 5.9
Circuses	13.88
Class(ification):	
– 0, 1, 2, 3 – see Surface spread of flame	
– of buildings (A, B, C)	3.1-3.4
– of fires A, B, C, D	10.2
Closely-seated audience – occupant capacity	5.19
Club concourse – occupant load factor	5.20
Colour blind people:	2.10
– sources of advice concerning	2.15
Commissionaire not acceptable as attendant	1.20
Common room – occupant load factor	5.20
Compartment, fire resistance of walls and doors	4.1
Competent person	1.6, 6.2, 11.5, 11.13
Complete blackouts in theatre auditoria	6.36
Complex, building enclosed within confines of	3.5
Conference room – occupant load factor	5.20
Continental seating	7.14
Conventional seating	7.13
Corridors:	5.46-5.51
– door opening into	5.95
– fire resistance of sub-dividing doors	4.1
– gradient of	5.50-5.51
– steps in	5.51
– sub-division of	5.49
– surface of floors of	5.88
Crush hall – occupant load factor	5.20

Cupboards on a protected route	5.47
Curtains	4.16-4.17
Dance floor – occupant load factor	5.20
Dead end, initial	5.14
Deaf people:	2.11-2.12
– fire alarms	2.12
– representatives of	2.3, 2.15
– sources of advice concerning	2.15
Decorations	4.19
– temporary	4.18
Devices, self-closing	5.102
Diagram 1	5.14
Diagram 2	5.27
Diagram 3a	5.42
Diagram 3b	5.43
Diagram 3c	5.43
Diagram 4a	5.65
Diagram 4b	5.65
Diagram 5	5.94
Diagram 6	7.3
Diagram 7	7.6
Diagram 8	7.7
Diagram 9	7.14
Diagram 10	7.19
Diagram 11	7.19
Diagram 12	7.19
Diagram 13	8.3
Diagram 14	11.8
Diagram 15	11.14
Diagram 16	11.14
Diagram 17	11.14
Diagram 18	11.27
Diagram 19	11.28
Diagram of electrical circuits	6.13
Diners' seating	7.15
Dining room – occupant load factor	5.20
Directional signs	5.111-5.118
Disabled Living Foundation	2.15
Disabled people:	1.13, Chapter 2
– code of practice for means of escape for	2.2, 2.5, 2.14
– consultation with representatives of	2.3
– evacuation procedure for using lifts	2.14
– not to be unnecessarily excluded	2.1
– occupant capacity of premises	5.18
– representatives of	2.3, 2.15
– sources of advice concerning	2.15
– unaccompanied	2.6
– use of lifts by for escape	2.14
Discotheques	Introduction, 1.4, 11.43
Distance:	
– from gangway to exits	7.3
– of travel	5.10
– measuring	5.13

Doors:
- automatic releases 5.103
- exit 5.99-5.101
- exit, maximum aggregate width of 5.26
- fastenings on 5.104-5.109
- fire resistance of 4.1
- all stairway enclosures 5.64
- landings to 5.96
- not openable in direction of exit 5.107-5.108
- notices on 5.119-5.123
- on escape routes 5.94-5.97
- opening into stairway or corridor 5.95
- other than exits 5.98
- revolving 5.26
- sills 5.97
- sliding 5.110
- sub-dividing a corridor 5.49
- thresholds 5.97
- to balconies, bridges and walkways 5.53
- to open easily and immediately 5.99

Drapes 4.16-4.17

Dressing room, fire resistance of doors 4.1

Dried grass, flowers and foliage 4.20-4.23

Electric signs 6.18, 6.46-6.50

Electrical:
- installations, inspection of 6.52
- wiring – general 6.19-6.26

Elevated podium deck, building on 3.5

Emergency lighting: 6.27, 6.31-6.42
- failure of 1.24
- rated duration of 1.24

Enclosed structures having a stage and stage area 13.78

Enclosure of stairways 5.58-5.64

Ends of rows to be aligned 7.11

Entertainments involving animals 13.88

Escalators 5.74

Escape lighting 6.43

Evacuation:
- general arrangements for 5.8
- lifts 2.14, 5.8, 5.89
- plan 1.1, 12.4
- procedure for disabled people using lifts 2.14, 5.8, 5.89
- time, maximum, meaning of 3.4

Exclusion and admission of the public 1.24, 6.59

Exit:
- and directional signs 5.111-5.118
- capacity 5.21-5.26
- door, maximum aggregate width of 5.26
- doors 5.99-5.101
- signs 5.116, 6.44-6.45
  - maintained internally illuminated 5.114-5.115
  - photo-luminescent type material not acceptable 6.45
- fire 5.117
- width, unit of 5.22

Exits:
- from dressing rooms 8.33, 8.35
- from flies and grids 8.33
- from painting rooms 8.33, 8.35
- from property stores 8.33, 8.35
- from scenery stores 8.33, 8.35
- from stages 8.33-8.34
- from wardrobes 8.33, 8.35
- in room etc where there is a stage 5.41
- in theatre auditorium 5.41
- number of 5.29-5.35
- roof 5.93
- siting of 5.41-5.44
- to be unlocked 1.5
- which cannot easily be seen 5.118
- width of 5.36-5.40
External stairway 5.78

Failure of:
- emergency lighting 6.59
- normal lighting 6.59
Fastenings on doors 5.104-5.109
Film studio – occupant load factor 5.20
Fire alarms:
- standards 9.2
- alternative 2.12
- automatic systems 9.19-9.21
- call points for 9.5
- for deaf people 2.11-2.12
- in areas to which public not admitted 9.12
- in multi-use premises 9.11
- manually operated 9.3-9.4
- to be distinctive 9.3, 9.9
- visual 9.14-9.17
Fire appliances, access for 3.5
Fire blankets 10.10
Fire drills 1.10, 1.11
Fire exit signs 5.117-5.118
Fire fighting equipment 10.12-10.13
Fire growth potential in unoccupied areas 4.10
Fire resistance of doors, floors and walls 4.1
Fire safety signs:
- consideration of colour perception when siting 2.10
- notices and graphic symbols 5.112
- photo-luminescent type material not normally acceptable 6.45
- self-luminous 5.113
Fire spread 4.8-4.10
Fire stopping 6.21
Fire ventilator 8.14
Firearms 11.45-11.46
Firefighter, theatre 1.21
Firefighting lifts 2.14
Firemen's switches 6.46-6.48
Flame illuminants 6.41

Floor coverings, advice on	4.11
Floors:	
– and tiers	5.87-5.88
– fire resistance of	4.1
– gradient of	5.87
– openings in	5.52
– surfaces of	5.88
Flowers, artificial and dried	4.20-4.23
Foam pits and mats	4.24-4.27
Fog machines	11.27-11.38
Foliage, artificial and dried	4.20-4.23
Forty-five degree (45°) rule for siting of exits	5.42-5.43
Foyer	5.63
Furniture and furnishings, advice on	4.12-4.15
Fuses, siting of	6.14
Gallery, technicians', construction of	8.28
Gangways:	
– surface of floors of	5.88
– to be of adequate width	7.10
Gas:	
– lamps, control of individual	6.42
– leak, action if suspected	1.24
– safety	6.1, 6.3
Gradient of:	
– corridor or passage	5.50-5.51
– of ramp	5.79
Grasses, dried	4.23
Grid, construction of	8.28
Guidance on scenery and properties	8.39-8.43
Guide to Design of Cavity Barriers & Fire Stops	4.9
Gymnastic foam pits and mats	4.24-4.27
Handrails	5.80-5.81, 7.12
Hangings	4.16-4.17
Haystack lantern light	8.8-8.13
Headroom of stairways, lobbies etc	5.69
Hearing-impaired people	2.11-2.12
– sources of advice concerning	2.15
Heating appliances	6.54-6.58, 12.57
High intensity lighting	6.51
Hoists	5.89-5.92
Home-made pyrotechnics prohibited	11.3
Hose reels	10.4
Indoor sports events, seating at	7.13-7.14, 7.20
Induction loop systems not acceptable as fire alarms	2.13
Initial dead end	5.14
Inner and access rooms	5.27-5.28
Inspection of electrical installations	6.52
Internally illuminated exit signs, maintained	5.114-5.115
Joint Committee on Mobility for the Disabled	2.15

Landings 5.70, 5.96
Lasers 6.51, 11.40-11.42
Lettering of fire safety signs and notices 5.117, 5.119
Lifts: 5.89-5.92
- evacuation 2.14, 5.8
- firefighting 2.14
- motor 5.91
- motor room, fire resistance of walls and doors 4.1
- shaft, wiring in 6.24
- use in evacuation 5.8, 5.89
- use of by disabled people for escape 2.14
- well, fire resistance of walls and doors 4.1
Lighting:
- emergency 1.24
- for special effects 6.51
- gallery, construction of 8.28
- normal 1.24
- switches, local, siting of 6.17
Limit on number of persons to be admitted 5.18
Liquefied petroleum gas (LPG) portable heaters 6.58
Lobbies:
- headroom of 5.69
- steps in 5.51
- surface of floors of 5.88
Lobby 5.63
Log book 1.5
Lounge - occupant load factor 5.20
LPG (Liquefied petroleum gas):
- containers and cylinders 6.58, 13.59, 13.88
- portable heaters 6.58, 12.57
Luminous discharge lighting 6.46-6.50

Main switches to be labelled 6.15
Maintained internally illuminated exit signs 5.114-5.115
Manager, responsibilities of 1.3-1.11
Manually operated fire alarms 9.3-9.4
Maximum aggregate width of exit door 5.26
Maysfield Leisure Centre 4.24
Means of escape:
- access and inner rooms 5.27-5.28
- accommodation of children 5.9
- accommodation stairways 5.68
- aggregate exit capacity 5.26
- aggregation of exits 5.26
- area of high fire risk 4.1-4.2, 5.27
- automatic door releases 5.103
- balconies 5.53
- barriers across escape routes 5.86
- barriers, protective 5.82-5.85
- basements, stairways to 5.66-5.67
- bridges 5.53
- buildings with only one escape stairway 5.56, 5.65
- calculated occupant capacity 5.19
- children separated from parents/guardians 1.17,1.19

- corridors: 5.46-5.51
  - door opening into 5.95
  - fire resistance of sub-dividing doors 4.1
  - gradient of 5.50-5.51
  - steps in 5.51
  - sub– division of 5.49
  - surface of floors of 5.88
- crush hall – occupant load factor 5.20
- cupboards on a protected route 5.47
- dead end, initial 5.14
- devices, self-closing 5.102
- directional and exit signs 5.111-5.118
- disabled people, occupant capacity of premises 5.18
- distance from gangway to exit 7.3
- distance of travel 5.10
  - measuring 5.13
- doors:
  - automatic releases 5.103
  - connecting directly with auditorium 5.64
  - exit 5.99-5.101
  - exit, maximum aggregate width of 5.26
  - fastenings on 5.104-5.109
  - landings to 5.96
  - not openable in direction of exit 5.107-5.108
  - notices on 5.119-5.123
  - on escape routes 5.94-5.99
  - opening into stairway or corridor 5.95, 5.103
  - other than exits 5.98
  - revolving 5.26
  - sills 5.97
  - sliding 5.110
  - sub-dividing a corridor 5.49
  - thresholds 5.97
  - to balconies, bridges and walkways 5.53
  - to open easily and immediately 5.99
- enclosure of stairways 5.58-5.64
- escalators 5.74
- escape lighting 6.43
- evacuation:
  - general arrangements for 5.8
  - lifts 2.14, 5.8, 5.89
  - plan 1.1, 12.4
  - procedure for disabled using lifts 2.14, 5.8, 5.89
  - time, maximum, meaning of 3.4
- exit:
  - and directional signs 5.111-5.118
  - capacity 5.21-5.26
  - door, maximum aggregate width of 5.26
  - doors 5.99-5.101
  - signs: 5.116, 6.44-6.45
    - maintained internally illuminated 5.114-5.115
    - photo-luminescent type material not acceptable 6.45
  - fire 5.117
  - width unit of 5.22

- exits:
  - from dressing rooms 8.33, 8.35
  - from flies and grids 8.33
  - from painting rooms 8.33, 8.35
  - from property stores 8.33, 8.35
  - from scenery stores 8.33, 8.35
  - from stages 8.33-8.34
  - from wardrobes 8.33, 8.35
  - in room etc where there is a stage 5.41
  - in theatre auditorium 5.41
  - number of 5.29-5.35
  - roof 5.93
  - siting of 5.41-5.44
  - to be unlocked 1.5
  - which cannot easily be seen 5.118
  - width of 5.36-5.40
- external stairway 5.78
- fastenings on doors 5.104-5.109
- fire exit signs 5.117
- fire safety signs:
  - consideration of colour perception when siting 2.10
  - notices and graphic symbols 5.112
  - photo-luminescent type material 6.45
  - self-luminous 5.113
- floors:
  - and tiers 5.87-5.88
  - gradient of 5.87
  - openings in 5.52
  - surface of 5.88
- forty-five degree rule for siting of exits 5.42-5.43
- foyer 5.63
- gangways:
  - surface of floors of 5.88
  - to be of adequate width 7.10
- general arrangements for evacuation 5.8
- general principles 5.5-5.6
- gradient of:
  - corridor or passage 5.50-5.51
  - ramp 5.79
- handrails 5.80-5.81, 7.12
- headroom of stairways, lobbies etc 5.69
- hoists 5.89-5.92
- initial dead end 5.14
- inner and access rooms 5.27-5.28
- internally illuminated exit signs, maintained 5.114-5.115
- landings 5.70, 5.96
- lettering of fire safety signs and notices 5.117, 5.123
- lifts: 5.89-5.91
  - evacuation 2.14
  - firefighting 2.14
  - motor 5.91
  - use in evacuation 5.8, 5.89
  - use of by disabled people for escape 2.14
- limit on number of persons to be admitted 5.18

- lobbies:
  - headroom of 5.69
  - steps in 5.51
  - surface of floors of 5.88
- maximum aggregate width of exit door 5.26
- measurement of width 5.39-5.40
- mirrors on escape routes 5.124
- notices on doors 5.119-5.123
- number of exits 5.29-5.35
- occupancy calculation: method 5.26
- occupant:
  - capacity 5.18-5.20, 7.8
  - load factor (m^2 per person)–Table C 5.20
- openings:
  - in floors 5.52
  - other than exits 5.98
- padlocks etc to be removed and kept in approved place 5.109
- painting rooms, exits from 8.35
- panic bolts and latches 5.106
- persons with impaired mobility 2.1, 2.4-2.9
- phased evacuations 9.6
- photo-luminescent type material 6.45
- protected lobby approach to single stairway building 5.65
- ramps 5.79
- revolving doors 5.26
- roof exits 5.93
- scenery stores, exits from 8.35
- self-closing devices 5.102
- self-luminous fire safety signs 5.113
- service roads used as 3.5
- signs, exit and directional 5.111-5.118
- siting of exits 5.41-5.44
- sliding doors: 5.110
  - notices on 5.110
  - to lifts 5.90
- small premises:
  - directional signs 12.39-12.46
  - doors 12.29-12.38
  - exit signs 12.39-12.46
  - exits 12.22-12.24
  - notices on doors 12.47-12.51
  - stairways 12.25
  - tiers 12.27
- Stage 1 travel 5.27-5.44
- Stage 2 travel 5.45-5.53
- Stage 3 travel 5.54-5.76
- stairways:
  - accommodation 5.68
  - construction of 5.70-5.73
  - directly serving a balcony or tier 5.62
  - door opening into 5.95
  - enclosure: prohibited contents 5.63
  - external 5.78
  - headroom of 5.69

– nosing of	5.73
– surface of treads of	5.88
– to basements	5.66-5.67
– ventilation of	5.77
– width of	5.37-5.38
– standing area – occupant load factor	5.20
– steps:	
– going of	5.72
– in corridors, passages and lobbies	5.51
– nosing of	5.73
– rise of	5.71
– surface of treads of	5.88
– sub-division of corridors	5.49
– Summerland fire	5.9
– surface finishes on escape route	4.3
– Table B – distance of travel	5.10
– Table C – occupant load factor	5.20
– Table D – exit capacity	5.26
– temporary structures:	
– distance of travel	13.40-13.41
– exit and directional signs	13.34-13.39, 13.49, 13.57
– exits	13.44-13.48
– means of escape – general	13.30-13.32
– occupant capacity	13.33
– ramps	13.41-13.42
– site selection and access	13.2-13.5
– tent ropes and pegs not to cause obstruction	13.48
– tiers	5.87-5.88
– turnstiles	5.26
– unit of exit width	5.22
– use of lifts by disabled people for escape	2.14
– vehicle routes used as	3.5
– ventilation of stairway	5.77
– ventilation systems	5.125
– vision panels	4.1, 5.27
– walkways	5.53
– wardrobes, exits from	8.35
– wheelchairs:	
– not to cause obstruction	2.8
– ramps for	2.9
– to be readily available	2.8
– users	2.4-2.9
– width:	
– measurement of	5.39-5.40
– of corridor	5.48
– of exits	5.36-5.40
– of ramp	5.79
– of stairways	5.37-5.38
Measurement of width	5.39-5.40
Minimum fire resistance – Table A	4.1
Mirrors on escape routes	5.124
National Federation of the Blind of the UK	2.15
Nitrogen vapour effects – safety	11.32

Non-licensed premises	1.1
Normal lighting	6.27-6.30, 6.43
Normal lighting, failure of	1.24
Notices on doors	5.119-5.123
Number of exits	5.29-5.35
Occupancy calculation: method	5.26
Occupant:	
– capacity	5.16-5.20
– capacity, calculated	7.8
– capacity, needs of disabled people to be considered	5.18
– load factor (m^2 per person) – Table C	5.20
Oil-burning heaters	6.57
Open stage	8.20-8.23
Openings:	
– in floors	5.52
– other than exits	5.98
Outdoor sports stadia, seating at	7.25, 7.27
Overcrowding	1.14
Padlocks etc to be removed and kept in approved place	5.109
Painting-rooms:	
– exits from	8.35
– separation of	8.36
Panic bolts and latches	5.106
Partition, surface of	4.7
Passages and passageways - see Corridors	
Pedestrian precinct, building in	3.5
People with impaired vision	2.10
Peripheral seating	7.16
Persons with impaired mobility	2.1, 2.4-2.9
Phased evacuations	9.6
Photo-luminescent type material	6.45
Piazza, building in	3.5
Pneumatic structures	13.21, 13.24-13.27
Podium deck, elevated, building on	3.5
Pop concerts:	1.16, 7.21-7.25, 8.2, 8.27
– gangways at	7.23
– seating at	7.21-7.25
– stages at	8.2, 8.27
– strobes	11.43
Portable fire extinguishers	10.5-10.9, 10.11-10.13
Premises:	
– forming part of a building	3.7
– managed by voluntary organisations	12.6
– not fully equipped as theatres, scenery for use in	8.53-8.56
– presenting special problems to licensing authority	3.6
Profoundly deaf people and fire alarms	2.12
Properties:	
– for use on open stage	8.21-8.23
– guidance on	8.39-8.43
Property stores:	
– exits from	8.35
– separation of	8.36
Protected lobby:	
– approach to single stairway building	5.65

– fire resistance of walls and doors 4.1
Protected route, fire resistance of walls and doors 4.1
Public address system as fire alarm 9.10
Pyrotechnics:
– ABTT Code of Practice for 11.8
– caution to be exercised in older buildings 11.7
– competent person to be in charge 11.5, 11.13
– firing boxes 11.8
– home-made, not to be used 11.3
– licensing authority to assess risks involved 11.7
– misfires 11.12
– safety factors 11.9-11.14
– seven days notice of proposal to use required 11.2
– siting of 11.6-11.7
– storage of 11.15-11.25
– untrained person to load or fire 11.13

Qualified person 1.7, 6.2
Queueing area occupant load factor 5.20

Radio studio – occupant load factor 5.20
Ramps 5.79
Rated duration of emergency lighting 1.24
RCD – see Residual current devices
Reading room – occupant load factor 5.20
Real flame, use of to be approved 11.39
Recording studio – occupant load factor 5.20
Records to be kept 1.5, 1.10
Residual current devices (RCD) 1.7, 6.6, 6.8-6.9
Restaurant – occupant load factor 5.20
Revolving doors 5.26
Roof exits 5.93
Royal Association for Disability and Rehabilitation 2.15
Royal National Institute for the Blind 2.15
Royal National Institute for the Deaf 2.15
Rubbish disposal 13.84

Safety check before public admitted 1.5
Safety curtain 8.4-8.6
Scenery:
– for use on open stage 8.21-8.22
– guidance on 8.39-8.43
– in premises where excess of permitted 8.56
– in premises where special effects permitted 8.56
– suspension of 8.29-8.32
– testing of 8.44-8.49
Scenery stores: 8.16-8.19, 8.36
– exits from 8.35
Scottish Council on Disability 2.15
Seating:
– alongside arenas, dance floors, skating rinks etc 7.16
– arrangements 7.17-7.20
– at indoor sports events 7.13-7.14, 7.20
– at outdoor sports stadia 7.25

– at pop concerts	7.18
– continental	7.14
– conventional	7.13
– dimensions	7.5
– diners'	7.15
– in boxes and enclosures	7.19
– lounge	7.15
– peripheral	7.16
– plan to be submitted for approval	7.1
– temporary	7.19
– to be fixed to the floor	7.17-7.19
Self-closing devices	5.102
Self-luminous fire safety signs	5.113
Separated stage:	8.2-8.19
– fire resistance of walls and doors	4.1
Service roads used as means of escape	3.5
Signs, exit and directional	5.111-5.118
Site selection – temporary structures	13.2-13.5
Siting of:	
– exits	5.41-5.44
– of local lighting switches	6.17
– of switchgear and fuses	6.14
Sliding doors:	5.110
– notices on	5.110, 12.51
– to lifts	5.90
Small premises:	
– application of Chapter 12	12.1
– attendants and duties	12.7-12.12
– basement/no basement	12.3
– calling the fire brigade	12.13
– construction and fire separation	12.14-12.20
– curtains, drapes, hangings etc	12.21
– directional signs	12.39-12.46
– doors	12.29-12.38
– electrical installations	12.52
– exit signs	12.39-12.46
– exits	12.22-12.24
– fire alarms	12.71
– fire fighting equipment	12.72
– floor coverings	12.21
– floors	12.28
– furniture and furnishings	12.21
– gangways	12.61-12.65
– gas installations	12.52
– heating	12.57
– licensee's responsibilities	12.4
– lighting	12.53-12.56
– location of audience accommodation	12.3
– managed by voluntary organisations	12.6
– manager's responsibilities	12.5-12.6
– matters not covered – see main index	
– maximum area of hall, room or auditorium	12.1
– maximum number of public present	12.1, 12.3
– no permanent provision for seated audience	12.1

– notices on doors 12.47-12.51
– oil installations 12.52
– seating 12.58-12.61
– stage and stage area 12.66-12.70
– stage in single-storey hall 12.1
– stairways 12.25
– tiers 12.28
– voluntary management committees 12.6
Smoke:
– and vapour effects 11.26-11.38
– machines 11.27, 11.33-11.38
– spread 4.8-4.9
Smoking:
– restriction of 1.22-1.23,13.14
– where prohibited 1.22, 1.23, Appendix B
Snooker room – occupant load factor 5.20
Socket outlets 6.7, 6.11
Space divider, surface of 4.7
Spacing between seating 7.6-7.7
Sports events, seating at 7.20-7.25
Staff:
– fire alarms 9.8
– fire instruction notices 1.12
– room – occupant load factor 5.20
– training 1.9-1.11
Stage 1 travel 5.27-5.44
Stage 2 travel 5.45-5.53
Stage 3 travel 5.54-5.76
Stage :
– construction of 8.24-8.28
– ventilation 8.7,8.14-8.15
Staged fire alarms 9.6-9.8
Stairways:
– accommodation 5.68
– construction of 5.70-5.73
– directly serving a balcony or tier 5.62
– door opening into 5.95
– enclosure: prohibited contents 5.63
– external 5.78
– fire resistance of walls and doors 4.1
– headroom of 5.69
– nosing of 5.73
– surface of treads of 5.88
– to basements 5.66-5.67
– ventilation of 5.77
– width of 5.37-5.38
Standing area – occupant load factor 5.20
Standing on seats 1.14
Stardust Disco Introduction, 4.10
Steps:
– going of 5.72
– nosing of 5.73
– rise of 5.71
– in corridors, passages and lobbies 5.51

- surface of treads of 5.88
Strobes (stroboscopic lighting) 6.51, 11.43-11.45
Structures, spacing between 13.6
Studio (film, radio, recording, tv) occupant load factor 5.20
Sub-division of corridors 5.49
Summerland fire 5.9
Surface:
- finishes on escape route 4.3
- of ceiling 4.6, Appendix A
- of wall 4.6, Appendix A
- spread of flame:
  - Class 0 definition 3.2, 4.4-4.6, Appendix A
  - Class 1/2/3 definition 3.3, 4.4-4.6, Appendix A
Suspension of scenery 8.29-8.32
Switches:
- lighting, local, siting 6.17
- main, to be labelled 6.15
Switchgear, siting of 6.14

Table A – minimum fire resistance 4.1
Table B – distance of travel 5.10
Table C – occupant load factor 5.20
Table D – exit capacity 5.26
Table E – minimum width of exits 12.24
Table F – minimum width of exits 13.35
Television studio – occupant load factor 5.20
Temporary decorations 4.18
Temporary seating 7.19
Temporary structures:
- access for fire brigade vehicles 13.2, 13.4
- access to hydrants and water supplies 13.5
- attendants and duties 13.9-13.13
- calling the fire brigade 13.15
- combustible material 13.83
- construction 13.16-13.28
- curtains, drapes, hangings etc 13.29
- distance of travel 13.40-13.41
- electrical safety 13.58
- exit and directional signs 13.49-13.57
- exits 13.34-13.39, 13.44-13.48
- fire alarm procedures 13.79
- fire fighting equipment 13.80
- fire spread 13.28
- flammable liquids, handling of 13.85, 13.87
- floor coverings 13.29
- floors and tiers 12.43
- furniture and furnishings 13.29
- gas safety 13.59
- grass and vegetation 13.82
- heat producing luminaires 13.67
- heating appliances 13.60-13.61
- licensee's responsibilities 13.1-13.5
- lighting 13.62-13.67
- LPG safety 13.59

- manager's responsibilities 13.7-13.8
- matters not covered – see main index
- means of escape – general 13.30-13.32
- occupant capacity 13.33-13.34
- platforms in 13.78
- ramps 13.41-13.42
- rubbish disposal 13.84
- seating and gangways 13.68-13.77
- site selection and access 13.2-13.5
- smoking restrictions 13.14
- spacing between (eg caravans) 13.6
- stages and stage areas in 13.78
- tent ropes and pegs not to cause obstruction 13.48
- use of lasers in 13.81
- use of pyrotechnics in 13.81
- use of strobes in 13.81
- vegetation 13.82

Tent ropes and pegs not to cause obstruction 13.48
Tented structures, materials 13.18-13.20
Testing of scenery 8.44-8.49
Theatre firefighter 1.21
Tiers 5.87-5.88
Timber Research & Development Association information sheet 4.1
Toilet, fire resistance of door to 4.1
Toilets, emergency lighting in 6.37
Town centre redevelopment scheme, building forming part of 3.5
Trailers, combustible materials not to be stored beneath 13.83
Travel distance - see Distance of travel
Turnstiles 5.26

Under-stage areas 8.37-8.38
Unit of exit width 5.22
Unoccupied areas, fire growth potential in 4.10
Use of batteries Appendix B
Use of lifts by disabled people for escape 2.14

Vapour effects 11.28-11.38
Vehicle routes used as means of escape 3.5
Vehicles, combustible materials not to be stored beneath 13.83
Ventilation:

- duct, fire resistance of floors and walls enclosing 4.1
- ducts, wiring in 6.24
- of stairway 5.77
- systems 5.125
- systems, wiring in 6.24

Vertical surfaces other than walls 4.7
Vision panels 4.1, 5.27
Vision-impaired people, sources of advice concerning 2.15
Voluntary management committees 12.6
Voluntary organisations managing premises 12.6

Waiting room – occupant load factor 5.20
Walkways 5.53

Walls:
- fire resistance of 4.1
- surface of 4.6, Appendix A
Wardrobes:
- exits from 8.35
- separation of 8.36
Water supply, failure of 1.24
Wheelchair users: 2.4-2.9
- access elsewhere than ground floor 2.7
- organised parties of 2.4
- provision of seating spaces for 2.5, 2.7, 7.1
- ramps for 2.9
Wheelchairs:
- not to cause obstruction 2.8
- to be readily available 2.8
Width:
- measurement of 5.39-5.40
- of corridor 5.48
- of exits 5.36-5.40
- of ramp 5.79
- of stairways 5.37-5.38

Printed in the United Kingdom for The Stationery Office
TJ002990 C12 11/00 10170